Environmental Footprints and Eco-design of Products and Processes

Series Editor

Subramanian Senthilkannan Muthu, Chief Sustainability Officer, Green Story Inc., Tung Chung, Hong Kong

Indexed by Scopus

This series aims to broadly cover all the aspects related to environmental assessment of products, development of environmental and ecological indicators and eco-design of various products and processes. Below are the areas fall under the aims and scope of this series, but not limited to: Environmental Life Cycle Assessment; Social Life Cycle Assessment; Organizational and Product Carbon Footprints; Ecological, Energy and Water Footprints; Life cycle costing; Environmental and sustainable indicators; Environmental impact assessment methods and tools; Eco-design (sustainable design) aspects and tools; Biodegradation studies; Recycling; Solid waste management; Environmental and social audits; Green Purchasing and tools; Product environmental footprints; Environmental management standards and regulations; Eco-labels; Green Claims and green washing; Assessment of sustainability aspects.

Aldo Alvarez-Risco ·
Subramanian Senthilkannan Muthu ·
Shyla Del-Aguila-Arcentales
Editors

Green International Business

Strategies, Research and Goals

Springer

Editors
Aldo Alvarez-Risco
Universidad Tecnológica del Perú
Lima, Peru

Subramanian Senthilkannan Muthu
SgT Group & API
Kowloon, Hong Kong

Shyla Del-Aguila-Arcentales
Escuela de Posgrado, Universidad San
Ignacio de Loyola
Lima, Peru

ISSN 2345-7651 ISSN 2345-766X (electronic)
Environmental Footprints and Eco-design of Products and Processes
ISBN 978-3-032-12496-8 ISBN 978-3-032-12497-5 (eBook)
https://doi.org/10.1007/978-3-032-12497-5

This Springer imprint is published by the registered company Springer Nature Switzerland AG
The registered company address is: Gewerbestrasse 11, 6330 Cham, Switzerland

Preface

The world is facing an unprecedented crossroads. Pressure on ecosystems, social fractures, and economic volatility demands a profound redefinition of international business practices. This book stems from the conviction that integrating sustainability into the core of business is not a concession, but the fundamental strategy for resilience and long-term success. The following chapters offer a multifaceted exploration of this transformation. We begin with the conceptual foundations of a sustainable economy (Chapter "Sustainable Economics: Uniting the Environmental, Social, and Economic Pluriverse"), understood as the inseparable convergence of environmental, social, and economic dimensions. We examine how global companies can translate this ideal into action, starting with the implementation of ethical practices in Human Resource Management (Chapter "Geopolitical Tensions and the Future of SDGs' Implementation in Latin America"). The central message is clear and hopeful: the international business sector has a unique power to contribute decisively to achieving the SDGs (Chapter "Win–Win Approach: How International Business Can Contribute to Meet the SDG"). Research IQ is also presented as a disruptive proposal for personal and institutional development in research (Chapter "Intangible Innovation: Non-Fungible Token (NFT) Price Evolution and Real-World Applications"). We demonstrate that sustainability drives value and innovation.

The analysis of the German "hidden champion" in Brazil (Chapter "Developing Global Environmental Management and Sustainability: The Case of a German Multinational "Hidden Champion" in Brazil") and the case of HP Inc. (Chapter "Sustainability-Driven Solutions Can Generate Business Growth—A Real-Life Example by HP Inc.") illustrate how sustainability-oriented solutions catalyze growth and competitive advantage. The Circular Economy (Chapter "Beyond Recycling: How the Circular Economy is Reshaping Business") is revealed as a transformative paradigm, beyond recycling, that is reshaping business models and global supply chains. This journey requires a profound change: cultivating a "Green Mindset" (Chapter "Green Mindset for International Business") that permeates the culture and decision-making of international businesses. Pioneering initiatives such as the Lima

2019 Green Sports Games (Chapter "Lima 2019: The First Green Sports Games") demonstrate the viability and impact of this comprehensive commitment.

This is a win–win path where economic value creation and positive impact reinforce each other. This work aims to serve as a practical guide for leaders, academics, and professionals dedicated to creating a future where business thrives in harmony with people and the planet. The urgency of the moment demands immediate and intelligent action.

Lima, Peru — Aldo Alvarez-Risco
Kowloon, Hong Kong — Subramanian Senthilkannan Muthu
Lima, Peru — Shyla Del-Aguila-Arcentales

About the Authors, Reviewers, Editors, and Their Contributions

This book would not have been possible without the incredible and dedicated efforts and contributions of its many authors, who wrote chapters and often assisted in reviewing the chapters of others.

Aldo Alvarez-Risco is coauthor of Chapter "Intangible Innovation: Non-Fungible Token (NFT) Price Evolution and Real-World Applications". Also, he is one of the editors and reviewers of the book.

Madeleine Bausch is coauthor of Chapter "Developing Global Environmental Management and Sustainability: The Case of a German Multinational "Hidden Champion" in Brazil".

Katyuska Borja-Paredes is coauthor of Chapter "Beyond Recycling: How the Circular Economy is Reshaping Business".

Shyla Del-Aguila-Arcentales is coauthor of Chapter "Intangible Innovation: Non-Fungible Token (NFT) Price Evolution and Real-World Applications". Also, she is one of the editors and reviewers of the book.

Sandra Huamán-Pastorelli is coauthor of Chapter "Win–Win Approach: How International Business Can Contribute to Meet the SDG".

Atanas Lipov is coauthor of Chapter "Sustainability-Driven Solutions Can Generate Business Growth—A Real-Life Example by HP Inc.".

Sandra Maycotte is coauthor of Chapter "Geopolitical Tensions and the Future of SDGs' Implementation in Latin America".

Mariano Naranjo-Bustios is coauthor of Chapter "Lima 2019: The First Green Sports Games".

A. Oarga-Mulec is coauthor of Chapter "Sustainable Economics: Uniting the Environmental, Social, and Economic Pluriverse".

K. R. Skene is coauthor of Chapter "Sustainable Economics: Uniting the Environmental, Social, and Economic Pluriverse".

Hristina Stoyanova is coauthor of Chapter "Sustainability-Driven Solutions Can Generate Business Growth—A Real-Life Example by HP Inc.".

Francisco Valderrey is coauthor of Chapter "Geopolitical Tensions and the Future of SDGs' Implementation in Latin America".

Emil Velinov is coauthor of Chapter "Green Mindset for International Business".

Contents

Editors and Contributors

About the Editors

Aldo Alvarez-Risco is a lecturer at Universidad Tecnológica del Perú. He obtained his Ph.D. from the Universidad Autónoma de Nuevo León in Mexico. Additionally, he holds a Doctor of Pharmacy and Biochemistry degree and a Master's degree in Pharmacology. He is a Pharmacist at the Universidad Nacional Mayor de San Marcos in Peru, as well as a Master's degree in Pharmaceutical Care from the Universidad de Granada in Spain. He has six years of experience in regulatory affairs at pharmaceutical companies and eight years of experience in the Ministry of Health in Peru. He has published more than 180 research publications, written some book chapters, and authored/edited books in the areas of Pharmaceutical Care, Sustainability, Entrepreneurship, and Circular Economy. He has been a lecturer in postgraduate programs since 2004 and a speaker at academic events in 22 countries.

Dr. Subramanian Senthilkannan Muthu currently works for SgT Group as Head of Sustainability and is based out of Hong Kong. He earned his Ph.D. from the Hong Kong Polytechnic University and is a renowned expert in the areas of Environmental Sustainability in Textiles and Clothing Supply Chain, Product Life Cycle Assessment (LCA), and Product Carbon Footprint Assessment (PCF) in various industrial sectors. He has five years of industrial experience in textile manufacturing, research and development and textile testing and over a decade's of experience in life cycle assessment (LCA), and carbon and ecological footprints assessment of various consumer products. He has published more than 100 research publications, written numerous book chapters, and authored/edited over 100 books in the areas of Carbon Footprint, Recycling, Environmental Assessment and Environmental Sustainability.

Shyla Del-Aguila-Arcentales is part of the staff at Escuela de Posgrado, Universidad San Ignacio de Loyola in Peru. Additionally, she holds a Doctor of Pharmacy and Biochemistry degree, as well as a Master's degree in Industrial Pharmacy from Universidad Nacional Mayor de San Marcos in Peru. Also, she is a pharmacist at the

Universidad Nacional de la Amazonia Peruana in Peru. She has 12 years of experience in the processes of pharmaceutical companies, including manufacturing, regulatory affairs, and quality control. She has published more than 135 research publications, written some book chapters, and authored/edited books in the areas of Pharmaceutical Care, Management, Smart Cities, Sustainability, Entrepreneurship, and Circular Economy. Also, she has been a lecturer in undergraduate and postgraduate programs since 2010.

Contributors

Aldo Alvarez-Risco Universidad Tecnológica del Perú, Lima, Perú

Katyuska Barja-Paredes University of York, York, England

Madeleine Bausch Department of Business Administration, Faculty of Economics and Business, University of Chile, Santiago, Chile

Shyla Del-Aguila-Arcentales Escuela de Posgrado, Universidad San Ignacio de Loyola, Lima, Perú

Sandra Huamán-Pastorelli Universidad Nacional del Callao, Callao, Perú

Atanas Lipov HP Inc., Sant Cugat del Valles, Spain

Sandra Maycotte Tecnologico de Monterrey, Monterrey, Mexico

Mariano Naranjo-Bustios Universidad Peruana de Ciencias Aplicadas, Santiago de Surco, Peru

A. Oarga-Mulec Materials Research Laboratory, University of Nova Gorica, Nova Gorica, Slovenia

K. R. Skene Biosphere Research Institute, Letham, Angus, UK;
The Living Laboratory, University of Dundee, Dundee, UK

Hristina Stoyanova Sofia University St. Kliment Ohridski, Sofia, Bulgaria

Francisco Valderrey Tecnologico de Monterrey, Monterrey, Mexico

Emil Velinov Newton University, Prague, Czech Republic;
RISEBA University of Applied Sciences, Riga, Latvia

Sustainable Economics: Uniting the Environmental, Social, and Economic Pluriverse

K. R. Skene and A. Oarga-Mulec

Abstract This chapter sets out to explore a theory of sustainable economics, grounded in the functionality and organization of the Earth system. It seeks to find a path to a sustainable future for humanity when viewed through all three lenses of environment, society, and economics. To begin with, the concept of a pluriverse is explained and critiqued. Next, the ecological pluriverse is explored, combining systems theory and thermodynamics. The Earth system consists of many worlds, or biomes, whose characteristics are emergent from the energetic contexts in which each biome finds itself. The biomes also operate their economies in ways that relate to their ecological properties. The totality results in an ecological pluriverse, one world and many worlds, setting the parameters for all life within. Next, the social pluriverse is explored, wherein a sustainable humanity can be seen to exist in resonance with the ecological pluriverse. Finally, the economic pluriverse is addressed, wherein appropriate sustainable economic activity emerges from the social pluriverse and, more fundamentally, the economics of the ecological pluriverse.

Keywords Localism · Resilience · Supply chains · Systems theory · Thermodynamics · Traditional knowledge

K. R. Skene (✉)
Biosphere Research Institute, Letham, Angus, UK
e-mail: krskene@biosri.org

The Living Laboratory, University of Dundee, Dundee, UK

A. Oarga-Mulec (✉)
Materials Research Laboratory, University of Nova Gorica, Nova Gorica, Slovenia
e-mail: andreea.oarga@ung.si

A. Alvarez-Risco et al. (eds.), *Green International Business*, Environmental Footprints and Eco-design of Products and Processes,
https://doi.org/10.1007/978-3-032-12497-5_1

1 Introduction

The Earth system and humanity find themselves in challenging times on many levels at present. Economics lies at the heart of both the problems and solutions, as our economic activities involve the use of the Earth system as both a source and a sink, through supply chains, extraction, and waste (Lampert 2019; Jie and Chen 2023). The existence of any species on Earth relies on them occupying available niche space in order to gain sufficient free energy to maintain internal order and remain alive within an entropic universe. Thus, our existence as a species is dependent on the Earth system, the overarching, interactive, emergent, nonlinear, suboptimal, feedback-driven, self-organizing complex collective of all physical, chemical, and biological interactions on the planet (Skene 2022a). The Earth system is organized into several biomes, including tropical rainforests, savannas, and tundras, each functioning as energetic subsystems. These biomes are connected through shared hydrological, atmospheric, and energetic global circulation systems, as well as the overarching cosmic laws of physics. Thus, the Earth system is both global and local in nature. Perturbation of the Earth system can lead to emergent, non-linear change, and niche availability can quickly be reduced or eliminated, leading to species extinction (Boennec et al. 2024). Thus, our hopes for a sustainable future for humanity rely upon the Earth system maintaining our niche space.

All multi-dimensional solutions require trade-offs (Farnsworth and Niklas 1995). Yet the dominant reductionist approach of much of the modern sustainability movement often fails to recognize this, due partly to a silo mentality, where territoriality, divisions between arts, humanities and sciences in formal educational and research establishments, a failure to recognize traditional knowledge, and a resistance to recognizing localism within globalism have all impacted upon our motivation and willingness to make these sacrifices, both economically and behaviorally (Skene 2021; Prescott 2023; Mathetsa et al. 2024). Gifford (2011) set out seven barriers to change, relating to climate destabilization, but applicable to the broader context of sustainability, which lead to inaction, or misplaced action: "limited cognition about the problem, ideological world views that tend to preclude pro-environmental attitudes and behavior, comparisons with key other people, sunk costs and behavioral momentum, discredence toward experts and authorities, perceived risks of change, and positive but inadequate behavior change." These barriers also represent areas where sustainability policy and presentation need to increase their efforts. Clearly, strong intellectual and emotional intelligence relating to the environment are key requirements.

The challenge in any multifactorial problem is to identify the unit of resolution and to have a consistent approach. Many of the key terms in sustainability are plastic words (sensu Poerksen 2010), whose meanings differ depending on the context and intention of their use, forming an ever-shifting virtual landscape, shrouded in a mist of confusion. Kirchherr et al. (2023) warn that such plasticity of meaning can lead to a situation where "multiple fragmented and often contested understandings may experience conceptual deadlock and eventual collapse." One example is the term "sustainability" itself, which is often separated into two broad schools: strong sustainability, where natural capital is maintained and critical natural capital is functionally and physically preserved, and weak sustainability, where natural capital is substitutable, and, provided total capital is maintained, then this represents a sustainable path (Neumayer 2003). Sustainability is used in two completely different and exclusive contexts in these cases. Biely and Chakori (2025, p. 63) have argued that "Despite the high dominance, and resistance, of the technocentric, growth-focused, development narrative, there is a promising body of work advancing scientific pluralism." In particular, they focus on social ecology.

Figure 1 explores some of the contributions and interconnections between the three arenas of environment, society and economics. Each influences the others and creates context for the path toward sustainability. However, key to this figure is the overarching Earth system, within which everything else is located. The Earth system itself is an outcome of thermodynamics and complex systems theory (Skene 2024). Fundamentally, everything else emerges within the Earth system, and this represents both an existential threat and an opportunity for life on Earth. Figure 2, based on work by Skene and Murray (2017), summarizes how the major schools of sustainability map onto these three arenas. As can be seen, each has a particular weighting toward one or more of the arenas. Again, none of these schools place the Earth system as the primary arbitrator and conveyor of sustainability (Young and Steffen 2009).

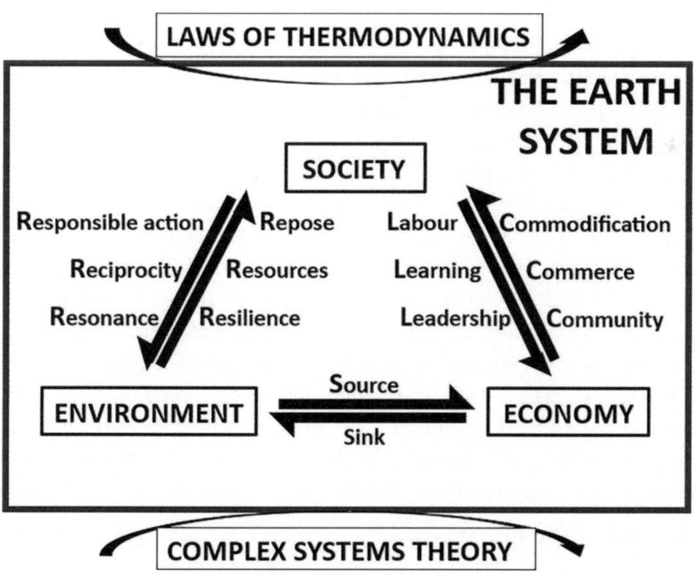

Fig. 1 Key interactions between human society, economics and the environment. All components are within the Earth system, which is an outcome of the laws of thermodynamics and complex systems theory

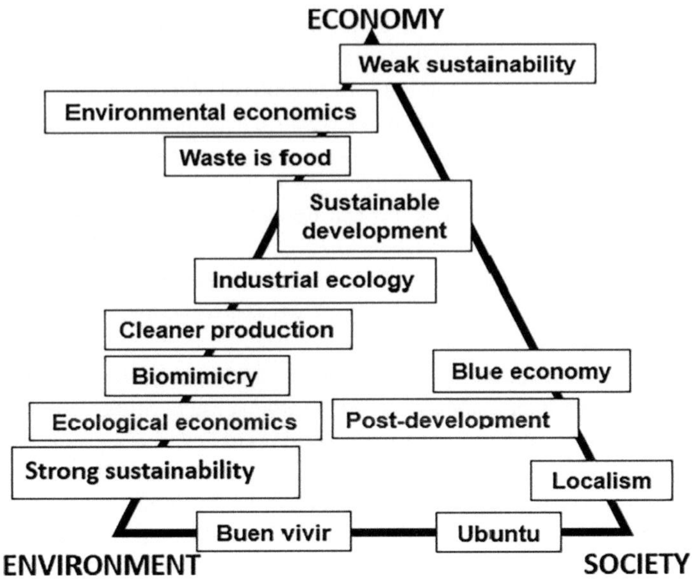

Fig. 2 Triangle of sustainability. The emphasis of major schools of sustainability in terms of the three pillars of sustainability (based on Skene and Murray 2017)

2 Chapter Outline

This chapter explores a theory of sustainable economics grounded in the functionality and organization of the Earth system. It is a ground-up approach, where a functional economy is built upon a functional society, and a functional society itself is built upon a functional ecology. All three aspects are within the Earth system. To begin with, the concept of a pluriverse is explained and critiqued. Next, the ecological pluriverse is explored, combining systems theory and thermodynamics. The Earth system consists of many worlds, or biomes, whose characteristics are emergent from the energetic contexts in which each biome finds itself. Each biome consists of nested layers of organization, each of which continually self-organizes within the greater context. The biomes also operate their economies in ways that relate to their ecological properties. Its totality results in an ecological pluriverse, setting the parameters for all life within. Next, the social pluriverse is explored, wherein a sustainable humanity can be seen to exist in resonance with the ecological pluriverse. Finally, the economic pluriverse is explored, wherein appropriate sustainable economic activity emerges from the social pluriverse and, more fundamentally, from the economics of the ecological pluriverse.

1. **The Concept of the Pluriverse**

The concept of a pluriverse was first referenced by James (1909, p. 125), who noted that: "Things are with one another in many ways, but nothing includes everything or dominates over everything." The emphasis here is on multiple realities rather than a single one, and multiple ontologies, or ways of knowing and being. These realities are often associated with cultures, particularly those with traditional knowledge, and these cultures are seen as emerging from a specific ecological context. Already, we can see that the pluriverse, in its broadest sense, straddles across society and environment. This is why it is the focus of this chapter, which seeks to find a path to a sustainable future for humanity when viewed through all three lenses of environment, society and economics.

Post-development, post-colonialism, and post-globalization literature have recently argued for considering additional, parallel paths for social and economic progress apart from the currently dominant Western Enlightenment thinking (see, for example, Escobar 2011; Savransky 2021). Skene and Oarga-Mulec (2024) write that "Different ontologies produce different realities, materialities, natures and worlds. Hence, our world is composed of a "pluriverse" of many different overlapping worlds with no clear boundaries or overarching principles that would turn this multiplicity into a singular world" (see also Blaser 2013; Escobar 2020). The concept of a pluriverse differs from a "universe" position, which emphasizes a single ontology. At the same time, the idea of a "multiverse" is an entirely different concept, grounded in astrophysics, which proposes that our universe, or cosmos, is merely one of many universes, which may even have different laws of physics, or share the same space as our universe, but in a different branch of the quantum wave function (Ellis 2011).

In many ways, the concept of the pluriverse, with one world and many worlds, resembles the global–local approaches (O'Riordan 2001; Quilley 2024) and much of the conversation relating to these is relevant here. The central challenge is to embrace both the unity and diversity of humanity and of the broader ecology. Chatterjee (1997, p 8) points out that "There cannot be just one modernity irrespective of geography, time, environment or social conditions. The forms of modernity will have to vary between different countries depending upon specific circumstances and social practices."

1. The Ecological Pluriverse

Of course, the Earth system, spanning its 4.5-billion-year history, has faced challenging times before. The Great Oxygenation Event, some 2.5 billion years ago, was probably the most significant pollution event in history, where the process of photosynthesis released the toxic gas, oxygen, into the oceans and atmosphere (toxic to the anaerobes, which represented almost all of life at that point). This led to one of the most significant mass extinction events in the planet's history (West 2022). By the Carboniferous, approximately 280 million years ago, atmospheric oxygen levels had reached levels of between 32 and 35% (Berner 2009; Kutschera and Niklas 2013). These high levels of oxygen resulted from the drawdown of carbon dioxide as vast forests grew and died, eventually forming the fossil fuels that would form the backbone of the industrial revolution and the current climate crisis, by releasing this carbon dioxide back into the atmosphere from whence it had been removed almost three hundred million years earlier, upon combustion (itself requiring oxygen).

Other significant extinction events included two "snowball Earth" episodes, around 700 and 650 million years ago, when much of the planet was covered in ice (Lu et al. 2022). These events ended due to extreme volcanic activity that increased atmospheric carbon dioxide levels, triggering a greenhouse effect. Additionally, at least five other mass extinction events occurred, eliminating up to 95% of life on Earth, primarily due to dramatic reductions in free energy availability (Skene 2015). Yet the Earth system has rebuilt and restored itself, different in form each time, but identical in function. This is because a function is a thermodynamic process with conservative outcomes. In contrast, forms are the outcomes of diffusion into ecological space, a thermodynamic process with a vast array of possibilities.

Skene and Oarga-Mulec (2024) note that the Earth system itself is a pluriverse. Each world, or biome, does not aspire to convert other worlds to their way of doing things, but rather, a whole series of biomes exist, fitting the energetic context. Due to the curvature of the Earth, incident radiation varies in density, highest at the equator and weakest at the poles. This establishes temperature and precipitation gradients, which govern the functionality and economics of biomes. In Table 1, we compare and contrast the characteristics of two such biomes, the Tundra/Boreal and Tropical Rainforests, in terms of their ecological, economic, and social organization.

It can clearly be seen that biomes differ considerably in terms of seasonality, evolutionary history, resilience/recovery economics, waste management and socio-ecological interactions. In terms of economic activity, carbon turnover times range from sixty-five years in tundra biomes to fourteen years in tropical forests (Carvalhais et al. 2014). Each biome is suited to the climate, seasonality and resource availability, both spatially and temporally, of the landscapes. While ice ages transformed the Tundra and Boreal biomes during the last 2.5 million years (the Quaternary), the Tropics were relatively unaffected.

It is clear that the biomes represent many worlds within one world, and the ecological pluriverse emerges from the physico-chemical conditions, both spatially and temporally. There is no conflict between biomes. For example, the desert biome is not forcing the temperate forest to align with its economic practices. Global dimensions are at play across all of the biomes, such as atmospheric gas content as well as global atmospheric and oceanic circulation patterns.

Universal laws of physics and chemistry operate across the entire Earth system, as does complex system theory, orchestrating resilience, recovery, self-organization

Table 1 A comparison of key ecological, economic and social characteristics of two biomes

Tundra/boreal	Tropical rainforest
Extreme seasonality	Stability and consistency
Recent history of significant climate change*	Little recent history of climate change*
Limited diversity	Huge diversity
Slow recovery	Rapid recovery
Storage economy	Spending economy
Slow recycling	Rapid recycling
Migration/transhumance	Establishment/settlement

*Over the last 2.5 million years

and temporal patterns. Yet even these are fine-tuned locally. Tectonic plate movement leads to changes in circulation patterns and movement of land and ocean, but the biomes remain, as they are fundamentally determined by differences in incident radiation density across the globe. What is also noteworthy in these two examples from Table 1 is that social and economic pluriverses emerge from this ecological pluriverse.

2. **The Social Pluriverse**

The role of society in economics can be traced back to the work of Adam Smith. He set out the foundations of free trade in his second book, An Inquiry into the Nature and Causes of the Wealth of Nations (Smith 1776), but his theory was built on an earlier book, The Theory of Moral Sentiments (Smith 1759). He focused on social responsibility (moral sentiment) and functionality as the "invisible hand" that would guide the free-market economy, often unintentionally (through self-interest), preventing excess and inequality. Skene (2022b) notes that "Unfortunately, the interconnected concepts within these two books, society and economics, became separated and the outcome has been an increase in inequality that has damaged society, resulting in a withered hand and the death of societal feedback." Zucman (2019) demonstrated that economic inequality has dramatically increased since the 1980s. Yet, more fundamentally, Smith failed to account for the Earth system, which had emerged as a concept around this time. Hutton (1788), the father of modern geology, compared the Earth to a superorganism, operating as a physiological unit. This is one of the earliest references to the planet as a system. The invisible hand of Smith (1776) imagined only two players in the theater, humans and the economy, but the more fundamental invisible fist, acting as a non-linear, emergent arbitrator across all economies on the planet, is the Earth system.

In indigenous societies, the concept of individual well-being is not only integrated within the community, but that community is integrated within its broader ecology and emergent from it, in such a way as to enhance and protect the life-giving capacities of an interactive, complex system as a whole (Baldwin 2013; Bignall et al. 2016). Hence, we can see strong linkages across all three arenas. Chibvongodze (2016, p. 157) writes that "The conservation of the environment in African societies rests upon a deliberate and conscious dovetailing of cultural and social identities with the natural and wildlife environment." Kuokkanen (2006, p. 269) reflected that sustainable thinking can be defined as "establishing and sustaining continuous relationships for the well-being of all."

Only with a pluriverse approach we can understand and maintain the local within the global, where the global represents the Earth system as a whole. The emphasis is on functionality rather than the possession of forms. Traditional knowledge does not recognize specialist, separate areas of knowledge. Philosophy, history, law, art, economics, science, anthropology, and sociology are all facets of the one complex reality. All is relational and integrated. These traditional cultures have emerged from the interaction of ecologically intelligent populations finding the global within the local. Traditional cultures may be understood as having found global patterns reflected in their local contexts, for example, indigenous agricultural practices that align with principles of sustainability.

There are universal traits, patterns, and cycles, emerging from the Earth system, that are recognized across the globe by humans who are integrated with that system through their landscapes, environments, and social and economic interactions. Furthermore, because many indigenous populations have lived through significant climate change events more rapid than the current crisis (e.g., the Younger Dryas ice age, during which it took just decades to reglaciate Northern Europe and decimate the forests of Scandinavia (Lothrop et al. 2011)), resulting in the need to relocate and migrate, they have the potential to be more resilient than Enlightenment populations in coping with the current environmental crisis (Kuecker and Hall 2011).

Thus, traditional knowledge, developed when humanity was in resonance with the Earth system, may provide direction, not in terms of actions and structures, but in terms of processes and foundations. The narrative intelligence locked up in many of the epic tales passed down as poetry, stories and song encapsulate so much of this knowledge, stretching back many thousands of years (e.g., Nunn and Reid 2016; Rahman et al. 2018; Brumm and Forth 2025). Stewart-Harawira (2005, p. xiv) claims that "far from [being] irrelevant in the modern world, traditional indigenous social, political and cosmological ontologies are profoundly important to the development of transformative alternative frameworks for global order and new ways of being." Their rootedness within the ecological pluriverse was not an option nor a convenience, but a necessity, as to live apart from the biomes in which they dwelt would have had existential consequences. The same could be said of our current crisis.

3. The Economic Pluriverse

All of life is pre-occupied by economics, which, at its most basic level, involves balancing the energetic accounts, ensuring sufficient free energy is available to do work, thus offsetting the losses accumulated from the second law of thermodynamics, the law of dissipation. It involves careful management of supply chains and of physiology, both in terms of growth, reproduction and maintenance. At higher levels of organization, temporally, ecosystem development (ecological succession) and population dynamics all have economic considerations. Without balancing the books, death awaits. We see increasing dissipation in all complex, self-organizing systems, achieved by increasing economic activity, provided free energy is available. This underpins, and is simultaneously directed by, many of the differences across the ecological and social pluriverse.

Schneiter and Bedi (2025) highlight that: "Policy implications indicate that economic growth alone may not be enough to lower emissions in all municipalities, highlighting the need for targeted strategies that account for local economic and environmental conditions." This is an interesting observation, hinting at the need to recognize an economic pluriverse embedded within the social and the environmental particularities of the locality. It should be added that historical elements, both socio-political and environmental, play a role here too, in terms of lag effects.

Povinelli (1995) recognized the significant issues surrounding globalism and economics, observing that "reconciliation with multiculturalism ends where a conceptual accommodation to a multi-economism would begin." Here, the social and the economic pluriverse can be seen to be co-dependent, and one without the other will fail. This concept of multi-economism has recently been explored by Holmes (2018). Indeed, modern geopolitical structures, such as competing states, the cult of the individual and market competition, are likely to add to the vulnerability of these "progressive" populations, rather than build resilience.

While reducing the complexity of supply chains increases resilience during trade wars, pandemics and conflicts, it is not merely the localizing of supply chains that is important. The ecological pluriverse must also be a core consideration, informing extraction, waste management and socio-economic practice. Depending on which biome we exist within, the way in which we manage resource processing should mirror the economic principles of that biome, as shown in Table 1. Only if our economics is embedded within the socio-ecological context can it usefully contribute to social and ecological sustainability, but also to economic sustainability. Horobet et al. (2024) recommend that regional economic solutions should recognize regional differences and that customized "strategies to expedite the transition in Western Europe and more specific interventions in Eastern Europe are essential to harmonize economic progress with environmental sustainability."

For example, the dominant politico-economic models still emphasize the centrality of economic growth in terms of delivering a better world, beginning with Kuznets (1955), who set forward the concept of economic growth driving a decrease in social inequality following an initial increase in inequality, in his classic U-shaped curve. Shafik and Bandyopadhyay (1992) extended this analysis to environmental damage, insisting that while at lower levels of economic growth there is an increase in environmental damage, increasing economic growth delivers an overall reduction in environmental damage.

However, these concepts have been challenged, with Stern (2003) concluding that "The evidence…shows that the statistical analysis, on which the environmental Kuznets curve is based, is not robust. There is little evidence for a common inverted U-shaped pathway that countries follow as their income rises." Jorgenson and Clark (2011), Dietz et al. (2012), Fujii and Managi (2013), Ravallion and Chen (2022), Kuźmar (2023), Altunöz (2023), Costa et al. (2024) and Wu et al. (2024) further challenge the assumption that social inequality and environmental damage will decline as a result of economic growth.

As mentioned earlier, a key property of any functional system is sub-optimality (trade-offs) across its components (Farnsworth and Niklas 1995). Wu et al. (2024), while exploring the importance of trade-offs globally, observed that: "while only a few countries simultaneously met the thresholds for prosperity and equality on the one hand and equality and environment on the other, none did for prosperity and environment." As a result, they emphasized the need for developed nations to prioritize environmental stewardship and economic redistribution over further growth.

Lynch and Long (2024) suggest that while the ecologically damaging consequences of capitalism may diminish with time, this is as a result of an aging dogma becoming stagnant and undergoing disintegration, rather than evidence of a positive relationship between economic growth and environmental and social sustainability (see also Van Rossem 1996; Matutinović et al. 2016). However, the displacement of one globalized economic theory by another is unlikely to lead to a sustainable economic future, given that the solutions need to be implemented across an ecological and social pluriverse, not an imagined universe, whose master is the human rather than the starker reality of the Earth system. Indeed, any economic theory that fails to be rooted in the ecological pluriverse will be unlikely to age well.

Given increasing strains that are a consequence of perturbing the Earth system, largely through our economic activities, and the impacts of emergent, non-linear system change on global supply chains, food production and stock markets, a shift to an economy embedded within the relevant biomes, with shorter supply chains and ecologically context-driven methods of production and waste management can only be a sensible path. Since resilience is a system characteristic, then sustainability can only be found in an integrated, relational approach. Thus, a sustainable economic program, and policies related to this, such as the post 2030 SDG replacements, should have at their core an understanding of the Earth system and a recognition of the essentiality of pluriversal thinking.

3 Conclusions

The importance of considering sustainability issues at the system level (in other words, having integrated policies that work across social, environmental and economic arenas), is still in its infancy. A reticence to recognize systemic change exists, and has its roots in the dominant reductionist economic and scientific philosophy of our time, the *Homo habilis* period of human history, wherein we are convinced that we can build a better world, rather than embedding ourselves into the actual reality of the Earth system, which pays no notice to whatever political or philosophical position we adopt; by their fruits shall we know them.

This chapter has explored the relationships between the three arenas of environment, society and economy, all emergent from the Earth system. However, for a sustainable future for humanity, it is essential that our economic activities are in resonance with the economy of the Earth system across the pluriverse of biomes that represent it across the globe. To do otherwise must surely be a strategy of doubtful value, as the Earth system is the arbitrator of our niche space and will respond to any perceived perturbation by restoring its functional identity, potentially removing the niche space that we inhabit. There is nothing personal in an extinction event. It is merely the consequence of systems theory and thermodynamics.

Many of the principles that allow sustainable living in one landscape are equally relevant to any other landscape, as there are global aspects such as thermodynamic laws, food chains, population dynamics, predator–prey relationships, hygiene and water management that cut across all landscapes. These global patterns and the interwoven nature of the Earth system, governed by the universal laws of physics, are expressed within a local context, and embrace both continuity and adventitiousness, the cyclical and the emergent. Resilience comes from how everything in Nature is connected and can shape-shift in an emergent way, conforming to the thermodynamic constraints but diffusing into available energetic space (Skene 2013, 2024).

However, this is not so much a "think globally, act locally" framework, but rather a resonance with the greater Earth system, which is expressed locally but shaped and driven at a global level. Fundamentally, this chapter suggests that indigenous philosophy and practice could be described rather as "think local, act local." Thinking begins at the local landscape level, producing appropriate spatial and temporal practice. The global is only recognized when comparative studies are made. What happens globally is reflected locally, so everything needed can be found locally, while what happens locally can be reflected globally, such as the cumulative effect of multiple local pollution events releasing greenhouse gases.

These global issues are important, and inform us at a different level, in terms of the affirmation of the Earth system as an entity, one world made of many worlds. The emphasis, however, is on locally derived wisdom and, as a result appropriate economic behavior.

The African philosopher, Okere (2005, p. 21) writes: "All philosophy is local and, even, individual before it can be universal; and nothing can be genuinely universally valid unless it was first authentically personal and inserted within a given culture. If this is the case for philosophy, it is likely to be the case for human knowledge, since every form of human knowledge must be situated or generated from within a culture or bounded by presuppositions, prejudgments, interests etc." This is indeed profound, expressing a much deeper understanding of the pluriverse and the importance of diversity, produced through the unfolded wealth of individual perception, which in turn requires us to be relationally empowered.

Geddes (1915, p. 392) writes that "Healthy life is completeness of relation of organism, function, and environment, and all at their best." Nature is not a limit to be overcome, neither is it raw in tooth and claw, but rather it is an emergent synergy, contextualizing our existence and requiring trade-offs, humility and gifting. To be one with nature is not some underdeveloped, primitive state of being, but rather, reflects true humanity, unfolded within this deeper context, embracing emergent complexity rather than isolated empiricism alone.

A philosophical pluriverse will be an essential concept as we move forwards within an existent ecological pluriverse, wherein systems theory and empiricism engage together. As García (2006, p.21), translated by Ezcurra (2023), stated, the Earth system is "a representation of a cut of complex reality, conceptualized as an organized totality, hence its denomination as a system, whose elements are not separable and therefore cannot be studied in isolation."

Acknowledgements We are extremely grateful to Dr Aldo Alvarez-Risco for the invitation to contribute to this volume. The authors acknowledge financial support from the Slovenian Research and Innovation Agency (research core funding no. P2-0412).

References

Altunöz U (2023) Analysis of Turkey's classic and financial Kuznets curves in regards to the level of development and increased distribution. J Ekon 5(2):60–69

Baldwin A (2013) Vital ecosystem security: emergence, circulation, and the biopolitical environmental citizen. Geoforum 45:52–61. https://doi.org/10.1016/j.geoforum.2012.01.002

Berner RA (2009) Phanerozoic atmospheric oxygen: new results using the geocarbsulf model. Am J Sci 309:603–606. https://doi.org/10.2475/07.2009.03

Biely K, Chakori S (2025) Sustainability transition theories: perpetuating or breaking with the status quo. Sustain Develop 33(1):52–68. https://doi.org/10.1002/sd.3101

Bignall S, Hemming S, Rigney D (2016) Three ecosophies for the Anthropocene: environmental governance, continental posthumanism and indigenous expressivism. Deleuze Stud 10(4):455–478. https://doi.org/10.3366/dls.2016.0239

Blaser M (2013) Ontological conflicts and the stories of people in spite of Europe. Curr Anthropol 54(5):547–567. https://doi.org/10.1086/672270

Boennec M, Dakos V, Devictor V (2024) Non-linearity and temporal variability are overlooked components of global vertebrate population dynamics. Diver Distrib 30(12):e13932. https://doi.org/10.1111/ddi.13932

Brumm A, Forth G (2025) Aboriginal flood narratives and the thunder complex in Southeast Asia. Archaeol Oceania 60(1):1–16. https://doi.org/10.1002/arco.5343

Carvalhais N, Forkel M, Khomik M, Bellarby J, Jung M, Migliavacca M, Saatchi S, Santoro M, Thurner M, Weber U, Ahrens B (2014) Global covariation of carbon turnover times with climate in terrestrial ecosystems. Nature 514(7521):213–217. https://doi.org/10.1038/nature13731

Chatterjee P (1997) Our modernity. Sephis

Chibvongodze DT (2016) Ubuntu is not only about the human! An analysis of the role of African philosophy and ethics in environment management. J Hum Ecol 53(2):157–166. https://doi.org/10.1080/09709274.2016.11906968

Costa EL, da Rocha FV, Péra TG, Caixeta-Filho JV (2024) Assessing the environmental Kuznets curve in agricultural transport: the case of Brazilian soy. J Agricult Appl Econ 56(4):666–691. https://doi.org/10.1017/aae.2024.31

Dietz T, Rosa EA, York R (2012) Environmentally efficient well-being: is there a Kuznets curve? Appl Geogr 32(1):21–28. https://doi.org/10.1016/j.apgeog.2010.10.011

Ellis GF (2011) Does the multiverse really exist? Sci Am 305(2):38–43. https://doi.org/10.1038/scientificamerican0811-38

Escobar A (2011) Sustainability: design for the pluriverse. Development 54:137–140. https://doi.org/10.1057/dev.2011.28

Escobar A (2020) Pluriversal politics: the real and the possible. Duke University Press, Durham

Ezcurra MP (2023) Beyond empiricism: Rolando García's theory of complex systems and the epistemological consequences of a nonlinear universe. In: Blanco MP, Page J (eds) Geopolitics, culture, and the scientific imaginary in Latin America. University of Florida Press, Gainesville, pp 305–322. https://doi.org/10.2307/j.ctvz938n9

Farnsworth KD, Niklas KJ (1995) Theories of optimization, form and function in branching architecture in plants. Funct Ecol 9(3):355–363. https://doi.org/10.2307/2389997

Fujii H, Managi S (2013) Which industry is greener? An empirical study of nine industries in OECD countries. Energy Policy 57:381–388. https://doi.org/10.1016/j.enpol.2013.02.011

García R (2006) Sistemas complejos: conceptos, método y fundamentación epistemológica de la investigación interdiscipinaria. Gedisa, Barcelona

Geddes P (1915) Cities in evolution: an introduction to the town planning movement and to the study of civics. Williams

Gifford R (2011) The dragons of inaction: psychological barriers that limit climate change mitigation and adaptation. Am Psychol 66(4):290. https://doi.org/10.1037/a0023566

Holmes C (2018) Polanyi in times of populism: vision and contradiction in the history of economic ideas. Routledge

Horobet A, Belascu L, Radulescu M, Balsalobre-Lorente D, Botoroga CA, Negreanu CC (2024) Exploring the nexus between greenhouse emissions, environmental degradation and green energy in Europe: a critique of the environmental Kuznets curve. Energies 17(20):5109. https://doi.org/10.3390/en17205109

Hutton J (1788) Theory of the earth. Trans Royal Soc Edinb 1:209–304

James W (1909) A pluralistic universe. Longmans

Jie H, Chen P (2023) Depletion of natural resources and environmental quality: prospects of energy use, energy imports, and economic growth hindrances. Resour Policy 86:104049. https://doi.org/10.1016/j.resourpol.2023.104049

Jorgenson A, Clark B (2011) Societies consuming nature: a panel study of the ecological footprints of nations 1960–2003. Soc Sci Res 40:226–244. https://doi.org/10.1016/j.ssresearch.2010.09.004

Kirchherr J, Yang NHN, Schulze-Spüntrup F, Heerink MJ, Hartley K (2023) Conceptualizing the circular economy (revisited): an analysis of 221 definitions. Resour Conser Recycl 194:107001. https://doi.org/10.1016/j.resconrec.2023.107001

Kuecker GD, Hall TD (2011) Resilience and community in the age of world-system collapse. Nat Cult 6(1):18–40. https://doi.org/10.3167/nc.2011.060102

Kuokkanen R (2006) The logic of the gift: reclaiming indigenous peoples' philosophies. In: Botz-Bornstein T, Hengelbrock J (eds) Re-ethnicizing the mind? Cultural revival in contemporary thought. Brill, pp 251–271

Kutschera U, Niklas KJ (2013) Metabolic scaling theory in plant biology and the three oxygen paradoxa of aerobic life. Theory Biosci 132:277–288. https://doi.org/10.1007/s12064-013-0194-3

Kuźmar S (2023) Income inequality and economic growth: the re-examination of the existence of Kuznets curve: evidence from CEE countries. Ekonomia i Prawo 22(2):237–250

Kuznets S (1955) Economic growth and income inequality. Am Econ Rev 45(1):1–28

Lampert A (2019) Over-exploitation of natural resources is followed by inevitable declines in economic growth and discount rate. Nat Commun 10:1419. https://doi.org/10.1038/s41467-019-09246-2

Lothrop JC, Newby PE, Spiess AE, Bradley JW (2011) Paleoindians and the younger Dryas in the New England-maritimes region. Quat Int 242(2):546–569. https://doi.org/10.1016/j.quaint.2011.04.015

Lu K, Mitchell RN, Yang C, Zhou JL, Wu LG, Wang XC, Li XH (2022) Widespread magmatic provinces at the onset of the Sturtian snowball Earth. Earth Planet Sci Lett 594:117736. https://doi.org/10.1016/j.epsl.2022.117736

Lynch MJ, Long MA (2024) Dissipative structure analysis and ecological disorganization: a critique drawn from ecological Marxist and treadmill-of-production approaches. World 5(3):551–569. https://doi.org/10.3390/world5030028

Mathetsa S, du Plooy C, Tayob K (2024) Breaking silo thinking within the South African water-energy-food nexus via systems thinking and simulation workshops. J Dig Food Energy Water Syst 5(1):146–158

Matutinović I, Salthe SN, Ulanowicz RE (2016) The mature stage of capitalist development: models, signs and policy implications. Struct Change Econ Dyn 39:17–30. https://doi.org/10.1016/j.strueco.2016.06.001

Neumayer E (2003) Weak versus strong sustainability: exploring the limits of two opposing paradigms. Edward Elgar, Northampton, MA

Nunn PD, Reid NJ (2016) Aboriginal memories of inundation of the Australian coast dating from more than 7000 years ago. Austr Geogr 47(1):11–47. https://doi.org/10.1080/00049182.2015.1077539

Okere T (2005) African philosophy and the hermeneutics of culture: essays in honour of Theophilus Okere, vol 2. LIT Verlag Münster

O'Riordan T (2001) Globalism, localism, and identity: fresh perspectives on the transition to sustainability. Earthscan

Poerksen U (2010) Plastic words: the tyranny of a modular language. Penn State Press

Povinelli EA (1995) Do rocks listen? The cultural politics of apprehending Australian aboriginal labor. Am Anthropol 97(3):505–518. https://doi.org/10.1525/aa.1995.97.3.02a00090
Prescott SL (2023) Planetary health requires tapestry thinking—overcoming Silo mentality. Challenges 14(1):10. https://doi.org/10.3390/challe14010010
Quilley S (2024) Schumacher against globalism and ecomodernism: ecology, subsidiarity and the politics of scale. Eur J Soc Theory 27(3):456–481. https://doi.org/10.1177/13684310241237428
Rahman A, Sakurai A, Munadi K (2018) The analysis of the development of the Smong story on the 1907 and 2004 Indian Ocean tsunamis in strengthening the Simeulue island community's resilience. Int J Disast Risk Reduct 29:13–23. https://doi.org/10.1016/j.ijdrr.2017.07.015
Ravallion M, Chen S (2022) Is that really a Kuznets curve? Turning points for income inequality in China. J Econ Inequal 20(4):749–776. https://doi.org/10.1007/s10888-022-09541-x
Savransky M (2021) Around the day in eighty worlds: politics of the pluriverse. Duke University Press, Durham, NC
Schneiter P, Bedi SM (2025) The environmental Kuznets curve revisited: a spatial panel model with heterogeneous coefficients. Energy Econ 143:108237. https://doi.org/10.1016/j.eneco.2025.108237
Shafik N, Bandyopadhyay D (1992) Economic growth and environmental quality: time series and cross-country evidence. In: Background paper for the world development report. World Bank, Washington DC
Skene KR (2013) The energetics of ecological succession: a logistic model of entropic output. Ecol Model 250:287–293. https://doi.org/10.1016/j.ecolmodel.2012.11.020
Skene KR (2015) Life's a gas: a thermodynamic theory of biological evolution. Entropy 17(8):5522–5548. https://doi.org/10.3390/e17085522
Skene KR (2021) No goal is an island: the implications of systems theory for the sustainable development goals. Environ Develop Sustain 23(7):9993–10012. https://doi.org/10.1007/s10668-020-01043-y
Skene KR (2022a) How can economics contribute to environmental and social sustainability? The significance of systems theory and the embedded economy. Front Sustain 3:980583. https://doi.org/10.3389/frsus.2022.980583
Skene KR (2022b) Steering the circular economy: a new role for Adam Smith's invisible hand. In: Stefanakis A, Nikolaou I (eds) Circular economy and sustainability. Elsevier, Amsterdam, pp 21–33. https://doi.org/10.1016/B978-0-12-819817-9.00028-4
Skene KR (2024) Systems theory, thermodynamics and life: integrated thinking across ecology, organization and biological evolution. Biosystems 236:105123. https://doi.org/10.1016/j.biosystems.2024.105123
Skene K, Murray A (2017) Sustainable economics: context, challenges and opportunities for the 21st-century practitioner. Routledge, Abington
Skene K, Oarga-Mulec A (2024) The circular economy: the butterfly diagram, systems theory and the economic pluriverse. J Circul Econ 2(3):9236
Smith A (1759) The theory of moral sentiments. Printed for Andrew Millar, in the Strand; and Alexander Kincaid and J. Bell, Edinburgh
Smith A (1776) An inquiry into the nature and causes of the wealth of nations. W. Strahan and T. Cadell, London
Stern DI (2003) The environmental Kuznets curve. In: International society for ecological economics internet encyclopedia of ecological economics.
Stewart-Harawira M (2005) The new imperial order: indigenous responses to globalization. Zed Books, London
Van Rossem R (1996) The world system paradigm as general theory of development: a cross-national test. Am Sociol Rev 61(3):508–527. https://doi.org/10.2307/2096362
West JB (2022) The strange history of atmospheric oxygen. Physiol Rep 10(6):e15214

Wu T, Rocha JC, Berry K, Chaigneau T, Hamann M, Lindkvist E, Qiu J, Schill C, Shepon A, Crépin AS, Folke C (2024) Triple bottom line or trilemma? Global trade-offs between prosperity, inequality, and the environment. World Develop 178:106595. https://doi.org/10.1016/j.worlddev.2024.106595

Young OR, Steffen W (2009) The earth system: sustaining planetary life-support systems. In: Folke C, Kofinas GP, Chapin FS (eds) Principles of ecosystem stewardship: resilience-based natural resource management in a changing world. Springer, Berlin, pp 295–315. https://doi.org/10.1007/978-0-387-73033-2_14

Zucman G (2019) Global wealth inequality. Annu. Rev Econ 11:109–138. https://doi.org/10.1146/annurev-economics-080218-025852

Geopolitical Tensions and the Future of SDGs' Implementation in Latin America

Francisco Valderrey and Sandra Maycotte

Abstract In an era of uncertainty, geopolitics and geoeconomics are becoming crucial tools for understanding international commerce and sustainable development. In this chapter, we conceptualize how those interconnected fields may reshape trade and business. The study examines the United Nations' Sustainable Development Goals (SDGs), with a focus on their implementation in Latin America. Specifically, we compare Mexico and Brazil, the two leading economies in the region, analyzing their local SDG implementation. As both countries' economies are significant recipients of Foreign Direct Investment (FDI) from China and the United States, we include this variable in our analysis, seeking external factors that influence business decisions. We conduct an overview of the implementation of the SDGs in Mexico and Brazil, as well as factors influencing business decisions on sustainability and the geopolitical alignment of Mexico and Brazil with China and the United States. We further discuss the perceived consequences of potential commercial alliances and their sustainability-related implications through scenario analysis, considering geopolitical factors. Thus, the study offers valuable insights into international business and sustainability, providing practical lessons for public and private decision-makers.

Keywords Geopolitics · International business · Latin America · SDGs · Sustainability

F. Valderrey (✉) · S. Maycotte (✉)
Tecnologico de Monterrey, Monterrey, Mexico
e-mail: francisco.valderrey@tec.mx

S. Maycotte
e-mail: sandra.maycotte@tec.mx

A. Alvarez-Risco et al. (eds.), *Green International Business*, Environmental Footprints and Eco-design of Products and Processes,
https://doi.org/10.1007/978-3-032-12497-5_2

1 Introduction

Recent events in the international arena are pointing to a profound transformation of the global order. The so-called "Pax Americana" has been shaken by unexpected conflicts and wars. By now, the rise of China is undeniable, surpassing the United States in many strategic areas while approaching the rival in many more. Economic leadership is arguably the cornerstone of American power, as well as military might, although distances are shortening between both nations. Technology is no longer under the monopoly of the United States, as evidenced by its widespread applications across various industries.

For almost a decade, the spotlight has been on the trade war between these superpowers; however, the commercial conflict is now expanding into other geographies. For instance, the BRICS are playing a significant role in international trade discussions, to potentially replace the US dollar in financial transactions, while seeking mutual trade agreements among member countries and future partner nations. Events in the Middle East are reshaping the entire region, extending conflict in Asia, while the war in Ukraine is testing the delicate equilibrium among Russia, the European Union, and the United States.

Despite the strategic importance of Asia and the European continent, a battleground is emerging between China and the United States that may tip the scales in favor of either side. Latin America, indeed, has become a silent battleground, providing the scenario for the advance of China into the region through commercial agreements, financial support, and FDI in the local economies. Americans have turned their interest to other latitudes, thus creating a void in a region that used to be considered their backyard. Presently, Latin America provides the ideal ground to compare and analyze the governance models offered by the most significant global economic powers, along with the benefits to local economies, political influence, and commercial and financial opportunities that come with alignment to either side (Dussel et al. 2024).

Latin America comprises a diverse array of more than twenty countries, and the number may nearly double when including the Caribbean. Nevertheless, few countries have considerable size and economic power, except for Mexico and Brazil, which maintain a dominant role. Historically, Central America has been much closer to Mexico and the United States. At the same time, South America has managed to escape the pressing influence of the United States by trading with Europe and Asia.

The rivalry between the two Latin American superpowers reaches trade and FDI (Correa da Cunha et al. 2024). The implications are numerous, as both nations compete for Chinese infrastructure projects, sizable energy investments, and technology transfers, while the United States offers opportunities in manufacturing, services, and market integration (Valderrey and Cabrera 2024). Although the money continues to flow to the region, the benefits are scattered and unevenly distributed, creating new forms of dependency (Yang and Paim 2024). Latin America is a major provider of commodities to global supply chains. Still, the negative externalities of mineral extraction and intensive agriculture often result in pollution, soil erosion, and even environmental disasters. Foreign MNEs in Latin America appear to exhibit a similar lack of respect for the local population.

We provide an analysis of how geopolitical tensions between China and the United States have a noticeable impact on the implementation of SDG policies in Latin America. We focus on Mexico and Brazil, as those two nations have the largest economies and populations, which leads to permanent competition for regional leadership. On the one hand, Mexico is fully integrated with North American markets, while Brazil has a diversified trade strategy, with political alignment with China and the Global South.

The chapter is divided into five sections, as follows. Following the introduction, we provide a conceptual review of geopolitics and geoeconomics, highlighting their impact on international business and trade. We emphasize the distinction between these two concepts, which are often used interchangeably, and explore future trends for geopolitical and geoeconomic analysis. In the next section, we analyze the experience of SDG policies and implementation in Mexico and Brazil. Next, we share our thoughts on the geopolitical alignment of the two Latin American nations under scrutiny, examining three different scenarios based on their proximity to the two external superpowers. In the final section, we present some conclusions and suggestions for future lines of research. In summary, this chapter examines the increasing significance of geopolitics and geoeconomics as key factors driving the emergence of new forms of globalization. At the center of the discussion is sustainability and how external forces shape the future of Latin America.

2 A Conceptual Review of Geopolitics and Geoeconomics

Although geopolitics and geoeconomics have been pivotal concepts for international relations and business, the distinction between the two terms frequently leads to misunderstanding. We will now explain both terms.

2.1 Differences Between Geopolitics and Geoeconomics

Geopolitics is closely tied to geography, encompassing the intersection of geographical space, natural resources, and spatial relationships within the context of international relations (Park 2023). Geopolitics emphasizes territoriality and national interests using politics, diplomacy, and military power. In a modern interpretation, geopolitics has expanded its original areas of interest to encompass multilateral institutions, regional alliances, and global governance (Hurrell 2024). On the other hand, geoeconomics examines the international arena through economic instruments, primarily related to commerce and trade. Such instruments are of interest to nations willing to use financial power in the pursuit of geopolitical ends (Sparke 2024).

Although the concept of geopolitics has been widely used for a long time, geoeconomics is gaining attention lately, as several nations have referred to their national security in terms of natural resources, as seen in the control of rare earth elements. When combining geopolitics and geoeconomics, there is a perfect match for claiming foreign territories, as the recent case of Greenland demonstrates. The United States has made clear its intention to obtain collaboration from the local inhabitants of the largest island in the Northern Hemisphere, at the expense of Denmark's sovereignty and the native population. China has placed similar claims in the South China Sea. Thus, military power may shape global outcomes (Eichengreen 2024).

The ongoing trade war between China and the United States has caused disruptions in the global supply chain. Two phenomena, clearly intertwined—dual circuit and nearshoring—have challenged well-established supply chains. Dual circuit refers to China's attempt to reduce its dependence on foreign trade, allocating a portion of its industrial output to merchandise exports while creating a second circuit, or loop, to provide the necessary products for local consumption. The COVID-19 pandemic highlighted how the Chinese government imposed export restrictions on goods considered strategically essential to combat the pandemic and secure raw materials, health supplies, and input products to maintain the country's manufacturing apparatus. On the contrary, nearshoring refers to the return of American enterprises to neighboring countries, specifically plants and manufacturing facilities, to reap the benefits of proximity to the world's largest consumer market. Those firms relocating in the geographical proximity obtained sizable cost reductions and customs savings. Furthermore, those companies were taken under the umbrella of the American government.

The commercial conflict has further implications, especially in the sustainable development of Latin America. China is undertaking infrastructure projects that are transforming the region (Bull 2024). The recent inauguration of the Chancay seaport in Peru has opened the door to the export of merchandise in quantities unseen before. A new shipping line was established to reduce cargo transportation time between China and Peru. Chancay is an example of how the Asian country's expansion into Latin America is fostering the so-called East Wing of the New Silk Road (Valderrey et al. 2020). The project, officially known as the Belt and Road Initiative (BRI), has been the most significant investment initiative involving logistics in recent years. Announcements of new projects come frequently, with many more to go in Latin America, as the land corridor connecting the Atlantic and Pacific Oceans crosses the Amazon.

The dark side of these projects lies in the negative externalities of construction, including excessive water usage, destruction of wildlife habitats, and depletion of natural resources that compete with human consumption. In theory, Chinese enterprises adhere to a detailed code of sustainability, but such a code is seldom fully respected. In addition to the environmental damage, easily visible throughout rainforests in the region, destruction also occurs to human populations living in the vicinity of mining exploitations or infrastructure developments. The effects go beyond resource allocation or health-related issues. As documented in the Observatory of Mining Conflicts in Latin America (https://www.ocmal.org/), some foreign enterprises are responsible for serious human rights violations. Therefore, it is imperative to balance the obvious economic benefits of FDI against the consequences of irresponsible business behavior.

2.2 *Trends in Geopolitical Analysis*

Geoeconomic and geopolitical analysis is undergoing significant changes. In addition to the convergence of geopolitics and geoeconomics, a notable trend is the widening of factors under scrutiny and the inclusion of non-traditional security threats. Cybersecurity is likely the primary concern, as hostile nations may develop technological superiority that could inflict severe damage on a rival country. It is well known how superpowers actively pursue the development of systems that can attack key systems overseas. Still, lately, new nations are also mastering technologies capable of blocking vital communication networks, government agencies, or financial centers, leaving large populations at the mercy of others. Geopolitics is highly sensitive to cybersecurity, and the case of Huawei was particularly significant due to its capabilities to collect private data through 5G hardware.

The menace from technology goes beyond cybersecurity. Currently, automobile companies such as Tesla and BYD are under investigation due to their ability to collect information from millions of consumers. Along with private date, the preference for a particular automobile brand may lead to dominance in emerging technologies, as some nations seek power through the creation of global standards that may rule the digital economy (Hsieh 2024), with profound implications for data governance, intellectual property and national security (Papa and Munier 2025).

In the military field, the challenge is significantly higher, as recent wars rely heavily on satellite capabilities to coordinate devices. It is well known how the Ukrainian forces were severely weakened when Starlink decided to limit the use of their proprietary internet network. There are numerous examples of how technology is increasing its relevance in military conflicts, as seen, for instance, during the Israel-Iran conflict. Presumably, technology will play a significant role in geopolitics as superpowers contend for space dominance.

Climate change is becoming one of those factors that were not considered part of traditional geopolitical threats. It is well known that the importance of rare earth minerals is well established, but more basic resources, such as water, are also at the center of geoeconomic ambitions (Birkenholtz 2023). Geographic locations prone to natural disasters are at the mercy of their neighbors, who have the capabilities to mitigate those hardships. Cases, such as the Mekong River's delta, provide a clear example of how China may inflict severe damage on other nations by controlling the downstream flow of water. Sustainability is gaining importance in the geoeconomic debate, due to increasing international regulations and the pursuit of the SDGs (Yang and Paim 2024).

Demography poses an increasing challenge to geopolitical and geoeconomic analysis. For instance, uneven wealth distribution may lead to migratory flows, with profound consequences for the host nations. Migration is at the center of the debate in highly developed countries, and some nations have leveraged it to gain geopolitical power, as exemplified by Turkey, which has facilitated migration between Europe and the Middle East. Declining birth rates pose serious concerns for South Korea and Japan, and many nations will likely not escape shrinking populations or changes in their citizenship composition. Along with demographics, health-related issues are also a cause for concern, as some countries may develop treatments for previously incurable diseases. It is well known how vaccine distribution was subordinated to geopolitical interests during the COVID-19 pandemic, to the point of labeling it vaccine diplomacy, which involves the supply of precious vaccines to allied countries.

In summary, a changing world demands new approaches to address geoeconomics and geopolitics, especially when considering the developments originating from artificial intelligence, which appears to be a game changer with unpredictable consequences.

3 SDG Adoption in Latin America: The Cases of Mexico and Brazil

A first glance into Latin America reveals an immense territory, characterized by uniformity, moderate cultural diversity, and similar stages of development across the region, among other similarities. The fact of the matter is that two nations stand out as the regional powers, disputing hegemony and resources from outside investors. Indeed, Mexico and Brazil are leading more than twenty countries, with no other countries producing a similar economic outcome, as they attract external investors by offering sizable internal consumer markets and territories.

3.1 A Comparative View of Mexico and Brazil

Mexico started the process of opening its economy after becoming a member of the OECD in 1994. Ever since, trade has been deeply associated with its northern neighbor, to the point of extreme dependence on bilateral trade. Despite Mexico's numerous attempts to diversify its exports through the signing of free trade agreements, the United States remains its largest trade partner. It is not only a matter of shipping products to the American market, but several industries are also profoundly interdependent. No industry offers a better example of interdependency than the automotive manufacturing sector.

Mexico has joined the top 10 list among automobile exporters, although a high percentage of the total value of car manufacturing locally is directly associated with foreign investors. Bordering states in northern Mexico attract numerous manufacturing investments due to lower wages and less stringent regulations. Many of those projects fall into the category of "maquila production", where the primary input from Mexico is provided by unskilled labor.

Many regions in Mexico, especially south of Mexico City, are no match for the Northern States, thus receiving much less investment. Those states have preserved rainforests and natural environments, although moderate industrial activity, tourism, and farming are compromising the last pristine ecosystems. Wealth concentrates in Mexico City, large cities, and the frontier states. In comparison to most countries in the region, Mexico boasts a moderately advanced economy, featuring an attractive domestic market and a robust export platform to the United States.

Brazil is the colossus of South America, both in terms of territory, population, and natural resources. Due to its economic self-reliance, decades of autarchic policies, leadership in specific strategic industries, and certain cultural factors, the country has maintained weak ties with Central and North America. The singularity of its language, Portuguese, in a majorly Spanish-speaking area of influence, keeps Brazil somehow isolated from neighboring nations. The industry and population are concentrated in the state of São Paulo, while most of the remaining citizens are spread along the Atlantic Ocean coastline. Nature is almighty in Brazil, with the Amazon basin occupying about two-thirds of the entire territory. Nevertheless, decades of mining and illegal farming activities are reducing national preservation areas and a massive part of the Amazon basin at an alarming rate.

Regarding trade, Brazil maintains a balance between export and import activities, spreading commerce evenly among the United States, China, the European Union, and neighboring countries, organized through the Mercosur free trade agreement. Although export revenues rely on a few commodities, the government has developed the aerospace industry, offshore energy exploration, bioenergy, and other sectors. The nation offers a diversified range of export products, while its internal market continues to attract foreign companies eager to sell to more than 200 million potential consumers. Table 1 presents a comparative view of Mexico and Brazil. There is an apparent difference between the two countries in terms of land area, population, economic output, and FDI attraction, favoring Brazil. Mexico, on the other hand, has a leading role in export figures, ease of doing business, and employment, with the rest of the indicators being almost even.

Mexico looks to the North, partnering with the United States and Canada, as the USMCA agreement protects trade and investment bonds. Brazil, instead, looks to the immediate vicinity, mainly Argentina, while keeping strong bonds outside the region. Geopolitics play a role, as Mexico aligns with its North American partners. At the same time, Brazil maintains a strong voice in international organizations, such as BRICS, while also holding a prominent position within the Global South nations.

Mexico and Brazil hold different views on sustainability. Essentially, Mexico adheres to the sustainability policies of foreign clients and investors, although it retains key decisions in those activities that are considered of strategic value. On the

Table 1 General indicators, Mexico and Brazil, 2023

Indicator	Mexico	Brazil
Total land area (sq km)	1,964,375	8,515,7767
Population (millions)	130	215
GDP (USD trillion)	1.46	2.08
Inflation rate (%)	4.8	4.5
Unemployment rate (%)	3.3	7.8
FDI (USD billion)	36	75
Exports (USD billion)	593	339
Global competitiveness index (score)	64.6	60.9
Ease of doing business rank	60	124

Source Own, with data from World Bank, UN SDG Database

contrary, Brazil claims the right to decide on these issues, demanding the opportunity to achieve a higher level of economic development before paying closer attention to SDG procedures. Large projects have been announced in either country, which will most likely further harm the environment. The geoeconomic interests of outsiders often overlook the importance of inclusive and sustainable development (Montiel et al. 2021). Table 2 presents a summary of key SDG indicators. There is little difference in per capita income, poverty rate, and the GINI Index results, which are primarily explained by income disparity. Uneven wealth distribution is a regional malady, with few countries finding a way out, even among those considered middle-income nations. Brazil outperforms Mexico in the race for renewable energy substitution and the attractiveness of FDI, while both countries are major pollutants.

For a long time, Latin America has faced this dilemma. It is a vast territory of immense natural wealth, although the abundance of resources has often acted more as a curse than a catalyst for development. Centuries of exploitation of mining resources and land for agricultural purposes have not depleted such wealth, although the rate of deterioration of the original natural endowment is becoming more evident.

Historical exploitation of resources by foreign powers has been a common occurrence in the region. Still, the race for dominance between China and the United States

Table 2 Key SDG indicators, Mexico and Brazil, 2024

Indicator	Mexico	Brazil
Human development index	0.758	0.754
GDP per capita (USD)	11,800	10,200
Poverty rate (% population)	36	32
Gini index (range 1–0)	45.4	48.9
Renewable energy (share, %)	24	46
FDI inflow (USD billions)	36	62
CO_2 emissions (Mt/year)	490	460

Source Own, with data from World Bank, UN SDG Database

may pose a new threat to Latin American countries. While sustainability is a significant issue in more advanced economies, not all nations in Latin America have made a full commitment to implementing the SDGs. Arguments such as a lack of financial resources or technological capabilities have limited public policies supporting the SDGs. At the same time, only MNEs and large local enterprises have taken visible actions in this regard. Income inequality, low levels of education, corruption, and violence are among the primary factors hindering the broader adoption of the SDGs. Despite the universal scope and nature of the SDGs, Latin American nations fall short of addressing climate change, inequality, or resource depletion, as progress in these issues requires collective action, which is hindered by local idiosyncrasies, traditions, and uneven power distribution. It is essential to recognize that the SDGs should be viewed in a broader context, extending far beyond environmental protection (Alvarez-Risco et al. 2021). Other challenges are directly associated with shifting power dynamics and institutional fragmentation (Ahmad et al. 2024).

On the positive side, the region has made progress in reducing extreme poverty, improving access to education, and expanding health services (Gómez-Pescador and Arzadum 2025). SDG compliance varies across nations, with some countries showing positive results, as seen in Chile or Costa Rica. Economic development and rules imposed by foreign corporations may have a positive impact, as many MNEs are subject to scrutiny from their headquarters.

4 Discussion

Latin America presents a fascinating backdrop for geopolitical and geoeconomic analysis, particularly when examining Mexico and Brazil. Table 3 presents potential outcomes for Mexico, and Table 4 presents potential outcomes for Brazil. Either table provides some elements for comparison with each nation, highlighting three alternative situations. In the first scenario, Mexico and Brazil choose a political alignment with either China. In the second scenario, both countries prefer an alignment with the United States. In the third scenario, Mexico and Brazil decide to seek equilibrium with both superpowers. Behind the scenes, an ongoing trade war is underway. No matter the temporary arrangements and announced agreements, both nations are clearly on a collision course.

4.1 Mexico: Geopolitics, Geoeconomics, and Sustainability

The geopolitical alignment of Mexico with China has been a concern for the Mexican government, both during the current and previous administrations. Former President López Obrador invited President Xi Jinping to an Inter-American reunion to bridge the Asian nation with Latin America. President Xi accepted the invitation, but soon he found a contradictory position from his Mexican counterpart. Surprisingly, López

Table 3 Geopolitical scenarios, Mexico

Decisions	China	US	Equilibrium
Infrastructure	Inter oceanic corridor	Railways	Road building
Trade	Free trade agreement	USMCA extension	Mediation in trade disputes
Logistics	Seaport construction	Liquid gas export seaport	Tren Maya
Local market	Eliminate tariffs	Eliminate taxes on digital services	Lower tariffs with China
Export platform	Manufacturing, US border	Export platform to third nations	Labor-intensive manufacturing

Source Modified from Valderrey and Cabrera (2024)

Table 4 Geopolitical scenarios, Brazil

Decisions	China	US	Equilibrium
Infrastructure	Financing infrastructure projects	Direct invitation to infrastructure projects	Amazon corridor
Trade	Extend BRICS	FTA with the United States	Bridging Mercosur and USMCA
Logistics	Connection with Chankay sea port	Seaport for American cargo vessels	Air cargo connection
Local market	Facilities for high-end technology products	Easy restrictions on digital services	Facilitate the import of consumer products
Export platform	BYD plant to service MERCOSUR	Manufacture production to re-export to Mercosur	Free trade zones in Manaus

Source Modified from Valderrey and Cabrera (2024)

Obrador shifted to a narrative of hostility toward China, which led to diplomatic repercussions and a lack of interest in carrying out several projects designed to increase China's presence in Mexico. Presently, this scenario appears unrealistic, primarily because the USMCA includes a clause that prohibits member nations (the United States, Mexico, and Canada) from entering into free trade agreements with non-market economies. Such a clause was explicitly designed to target China, as the Asian country has not yet achieved the status of a free-market economy. Nevertheless, erratic decisions from the current US administration may create unexpected situations. Therefore, we cannot rule out such a possibility, in which case, Mexico could adhere to the BRI initiative and become a candidate for sizable loans, state-of-the-art technology, oil exports, and job creation through manufacturing facilities intended to serve the US market. Major decisions could involve a substantial investment in infrastructure along the Inter-Oceanic Corridor, an ongoing project designed to compete with the Panama Canal. Trade decisions could impact the signing of a free trade agreement. In logistics, China could replicate the Chankay seaport in Peru and construct a large container seaport on the Mexican coastline, likely in the Baja California peninsula. Mexico could also favor the entrance of Chinese-manufactured products by eliminating tariffs or even fostering investments in manufacturing facilities along the US border, as an export platform.

The second scenario, aligning with the United States, essentially proposes maintaining the status quo. Indeed, Mexican trade and investment actions are geared toward this possibility, regardless of the ideological preferences of President Claudia Sheinbaum. Strategic decisions under this scenario could involve building railway infrastructure for merchandise and people, as well as extending the USMCA free trade agreement. Mexico could also offer full support to build export platforms for liquefied natural gas and other energy exports to Asian markets, eliminate local taxes on digital service providers, and redirect manufactured products from American enterprises into Central American markets and the Caribbean.

The final scenario involves a more cautious geopolitical involvement, reaping the benefits of trade with favorable terms with both superpowers, with an open eye to attract FDI. Strategic decisions will focus on road building, mediation in trade disputes, maintaining tariffs on Chinese imports at agreeable levels, and exploring labor-intensive manufacturing opportunities with Chinese enterprises for export to Latin American markets.

All three scenarios may have a profound impact on sustainability. In the first case, Chinese infrastructure projects may risk environmental damage if construction firms fail to follow established protocols. An increase in trade may have a positive impact, for instance, by utilizing advanced technologies in agriculture, while building a new seaport may compromise fragile coastal territories. Overall, the scenario looks pessimistic, with the potential for high emissions. In the second scenario, Mexico could benefit by applying much stricter rules for compliance with American requirements, as will probably happen in merchandise transportation and the transition to clean energies. Overall, this scenario may hinder local companies but benefit the population with better sustainability rules and regulations. In the third scenario, the lack of clear rules and the desire to please both superpowers may likely lead to increased pollution, consumption without proper limits, and a prioritization of economic interests over sustainability. Overall, this is not a desirable option for the environment.

4.2 Brazil: Geopolitics, Geoeconomics, and Sustainability

Like Mexico, Brazil may be compelled to make decisions, despite its traditional stance of neutrality and its aspiration to play a leading role in global forums. President Lula da Silva is inclined to foster friendly relations with nations of the Global South, where he feels more comfortable. He reacted to many strategic decisions in international relations from his predecessor, who was a close ally of the United States. Former president Bolsonaro is probably out of any political race, as well as his political supporters. Nevertheless, Bolsonaro still enjoys a fair degree of popularity and could get support from the army. He was a high-ranking military officer, and

Brazil has a long tradition of military intervention in politics. Approaching China is probably the status quo option, as the bilateral relationship is presently at its highest. Lula da Silva stops short of making major political declarations that could infuriate the United States, but continues to work toward maintaining a distance with President Trump. He can negotiate better commercial terms with the European Union, seek financial resources from China, while dealing with the United States under more favorable conditions to avoid confrontation.

In terms of strategic decisions, in this first scenario, Brazil could seek financial backing for infrastructure projects and extend free trade among BRICS countries, including both the original members and those newly admitted to the organization. In terms of logistics, it is essential to connect the Chankay seaport in Peru with the Santos seaport on the Atlantic Ocean. As for local market actions, Brazil could facilitate the distribution of Chinese high-end electronic products and pressure BYD to establish several plants to service MERCOSUR markets.

The second scenario appears less realistic, as Lula da Silva has little intention of following orders from President Trump. This scenario could become a reality if, for any reason, China were to turn its back on Brazil. Under those circumstances, Brazil could allow easy access to American construction companies, propose a free trade agreement with the United States, and build a seaport for American cargo vessels crossing the Atlantic from the East Coast. Lula da Silva could avoid restrictions on digital platforms and services from American enterprises while attracting partners to establish manufacturing facilities oriented toward serving Mercosur markets.

The equilibrium alternative could include the opening of the Amazon corridor to companies from both superpowers, bridging the MERCOSUR and USMCA markets, facilitating logistics operations with air cargo, and the import of consumer products. Additionally, Brazil could attract investments from the superpowers to expand manufacturing in the Manaus Free Trade Zone, situated in the Amazon region.

As for sustainability, the first option may bring a mixed impact to the environment. Chinese companies could likely manufacture and distribute solar panels and green energy products; however, on the negative side, logistics could harm the rainforests. Overall, this scenario could have both negative and positive impacts, although mining extraction and extensive agricultural usage could cause a more negative effect. The second option, instead, could result in the expansion of SDGs, as many American companies follow some strict protocols, even when operating overseas. High-technology acquisition could likely optimize industrial processes, reducing pollution and waste. This option appears to be more environmentally friendly. The third option may also bring positive outcomes in terms of sustainability. By opening itself to international standards and protocols, Brazil could likely be pressured to adhere to the SDG protocols. This option may yield the desired results for sustainable development.

5 Conclusions

Latin America presents an excellent opportunity for examining the impact of geopolitics and geoeconomics on sustainability. The region is under pressure to take sides with China or the United States, with two major players leading the game: Mexico and Brazil. Both superpowers have much at stake, although China seems to have a head start. The Asian nation has increased trade, favored significant investments in infrastructure projects, and offered generous financial deals. Additionally, China keeps close ties with local governments, regardless of corruption or lack of democratic values. The current situation is probably a consequence of the abandonment by the United States, as American attention focuses on the Middle East, Iraq, and Afghanistan, while losing touch with Latin American nations.

In this chapter, we provide a comparative analysis of two nations that are iconic in their management of international relations in diverse situations. While Mexico is a direct neighbor to the most powerful nation in the world, Brazil can manage its path of economic development and its foreign relations with other countries. As the world becomes increasingly polarized, the stories of these two countries may offer valuable lessons to national and local authorities, enterprises, and individuals who will likely face difficult choices. Those choices will almost inevitably impact the environment, and the SDGs will likely gain importance.

There are many opportunities for further investigation. A comparison should examine how firms respond to the described changes, how local management can benefit from understanding SDG procedures, and how geopolitical and geoeconomic factors integrate into the sustainability equation. A great deal of detailed information is required at the regional level.

References

Ahmad T, Saeed S, Kukreti M (2024) Shifting priorities: global geopolitical events and the EU's foreign policy approach to sustainable development. J Clim Commun Develop 3(2):53–68

Alvarez-Risco A, Del-Aguila-Arcentales S, Rosen MA, García-Ibarra V, Maycotte-Felkel S, Martínez-Toro GM (2021) Expectations and interests of university students in COVID-19 times about sustainable development goals: evidence from Colombia, Ecuador, Mexico, and Peru. Sustainability 13(6):3306

Birkenholtz T (2023) Geographies of big water infrastructure: contemporary insights and future research opportunities. Geogr Compass 17:8

Bull B (2024) China and the new geopolitics of climate multilateralism in Latin America. Iberoam Nordic J Latin Am Caribbean Stud 53(1):44–58

Correa da Cunha H, Singh V, Amal M (2024) Geopolitical risk distance and foreign direct investment in Latin America

Dussel E, Cook JA, Alter JS (eds) (2024) Connecting China, Latin America, and the Caribbean: infrastructure and everyday life. University of Pittsburgh Press

Eichengreen B (2024) Geopolitics and the global economy. J Int Money Finan 146:103124

Gómez-Pescador I, Arzadun P (2025) Sustainable development goals (SDGs) through business models: a bibliometric and content analysis. J Iberoam Acad Manag 15:11542

Hsieh PL (2024) Shaping green regionalism: new trade law approaches to environmental sustainability. Rev Eur Comp Int Environ Law 33(2):172–182
Hurrell A (2024) Geopolitics and global economic governance. Oxf Rev Econ Policy 40(2):220–233
Montiel I, Cuervo-Cazurra A, Park J, Antolín-López R, Husted BW (2021) Implementing the United Nations' sustainable development goals in international business. J Int Bus Stud 52(5):999
Papa J, Munier N (2025) The role of critical minerals demand in advancing the sustainable development goals (SDGs) in Latin America. Sustain Develop 15:242
Park J (2023) Rethinking geopolitics: geography as an aid to statecraft (online). Texas Natl Sec Rev 9:543
Sparke M (2024) Defining geoeconomics amid shifts in global hegemony: critical geographies of new international conjunctures. Environ Plan A Econ Space 41:5293
Valderrey FJ, Montoya MÁ, Sánchez A (2020) Latin America: the east wing of the new silk road. Compet Regul Netw Ind 14:1783591719853537
Valderrey F, & Cabrera, O (2024) Impacto del nearshoring en las relaciones México–China. In: Dussel E (ed) América Latina y el Caribe y China. Economía, Comercio e Inversión. UNAM
Yang CH, Paim MA (2024) Evaluating China-Brazil bilateral engagements in climate change and energy: a true sustainability partnership or green appropriation? Camb Int Law J 13(2):300–326

Win–Win Approach: How International Business Can Contribute to Meet the SDG

Sandra Huamán-Pastorelli

Abstract The SDGs offer a comprehensive framework for the private sector to align its strategies with global sustainability goals. However, companies face challenges prioritizing short-term objectives over the interconnected dimensions of the SDGs: economic, social, and environmental. The pandemic exacerbated economic inequalities, underscoring the need to strike a balance between market stability and fair working conditions. Multinational corporations, through sustainable business models and value chains, can advance the SDGs, although obstacles such as "SDG washing" and a lack of genuine commitment persist. Key sectors demonstrate progress: sustainable certifications in textiles, green energy technologies, and circular waste management. Collaborative approaches—such as public–private partnerships and win–win strategies—are crucial for linking innovation, equity, and environmental protection. Education and strong regulatory frameworks (SDGs 16–17) facilitate this transition, ensuring that economic growth does not compromise social well-being or planetary boundaries.

Keywords SDGs · Sustainable development goals · Business sector · Triple impact · Sustainability

1 Introduction

On September 25, 2015, 195 nations committed to the 17 Sustainable Development Goals (SDGs) and 169 targets as part of the 2030 Agenda for Sustainable Development (United Nations Development Programme 2020), with the private sector playing a significant role in the process (Scheyvens et al. 2016). However, the inherent complexity of sustainable development (Rittel and Webber 1973) poses challenges for businesses, which often prioritize short-term, smaller-scale objectives over broader, long-term sustainability goals (Montiel et al. 2020). In this regard, the

S. Huamán-Pastorelli (✉)
Universidad Nacional del Callao, Callao, Perú
e-mail: sehuamanp@unac.edu.pe

A. Alvarez-Risco et al. (eds.), *Green International Business*, Environmental Footprints and Eco-design of Products and Processes,
https://doi.org/10.1007/978-3-032-12497-5_3

SDGs can offer a structured framework to help companies align their strategies and operations with specific, targeted outcomes (Sullivan et al. 2018). The SDGs are holistic and inseparable, aiming to balance the economic, environmental, and social pillars of sustainable development. They are also deeply interconnected, relying heavily on effective governance and the active involvement of global stakeholders for their successful implementation (Maher and Buhmann 2019).

The SDGs economic dimension group includes 8, 9, 10, 12, and 17 goals. However, the disruptions caused by COVID-19 have led the global economy to experience growth following, but this recovery remains subdued compared to pre-pandemic levels. Labor market volatility was particularly pronounced in both developing and developed countries, suggesting that their economic recovery may be prolonged compared to regions less impacted by the crisis. Labor and economic growth are interdependent; the valorization of both is crucial for sustainable development. However, the current economic model prioritizes market stability and price controls, often at the expense of labor conditions and fair wages (Arte et al. 2023; Pilgrim and Bohnet-Joschko 2022).

The SDGs social dimension group includes Goals 1, 2, 3, 4, 5, 7, 11, 16, and 17. According to the World Trade Organization (2018), integrating the SDGs into business strategies can drive socially sustainable outcomes, including inclusive growth, job creation, and poverty reduction. Organizations with significant public influence prioritize social initiatives within their business frameworks and financing structures. Many companies have redefined their stakeholder engagement practices to align their objectives with those that benefit both their interests and broader societal goals. The SDGs' environmental dimension group includes 6, 13, 14, 15, and 17, which correlate with all dimensions. These goals are closely tied to the environmental performance of companies that implement eco-friendly practices in their operations. Promoting greater environmental awareness should begin within the educational system by integrating curricula (Huaman-Pastorelli 2020) focused on sustainable development and best practices in international trade. This approach equips future generations with the skills to implement these principles effectively (Arana et al. 2020; Huaman-Pastorelli 2020).

From another perspective, Montiel et al. (2020) categorize the 17 SDGs into six groups based on their impact on either positive or negative externalities, such as knowledge, wealth, and health, or overuse of natural resources, harm to social cohesion, and overconsumption. This classification helps companies align their strategies with broader societal objectives. By embedding SDGs into the Value Chain framework, specific goals can be connected to activities spanning supply, production, distribution, use, and disposal phases. This approach highlights how multinational corporations can integrate sustainability across their operations. Nowadays, the win–win approach (Shonk 2025) integrates economic, social, and environmental dimensions, aiming to cultivate entrepreneurial capacity that drives social sustainability. It calls for responsible corporate producers to support underdeveloped but resource-rich regions, promoting growth that prioritizes community well-being. A win–win approach is a negotiation strategy that aims to benefit all parties involved.

It's also known as integrative bargaining, cooperative problem-solving, or mutual gains negotiation.

It differs from other approaches, like win–lose or lose–lose, which assume that one party's gain is another's loss. Business negotiators recognize that achieving a win–win outcome is crucial; when both parties are happy with the agreement, the likelihood of establishing a durable and successful business relationship significantly increases. However, specific strategies for creating a win–win negotiation contract often appear difficult to pinpoint (Shonk 2025). The involvement of international business corporations in achieving the SDGs is rooted in the principle that everyone can benefit. When both parties are content with an agreement, the chances of long-term success increase significantly, as this typically leads to both sides willingly accepting the proposed solution. For example, a buyer and seller might agree on flexible payment terms, allowing the buyer greater financial flexibility while ensuring the seller retains a loyal customer. Similarly, in procurement, both buyers and suppliers gain advantages from their transactions.

2 Origin and Definitions

According to Sreenivasan and Suresh (2023), sustainable entrepreneurship research has strongly impacted SDG 11 (Sustainable Cities and Communities), SDG 12 (Responsible Consumption and Production), and SDG 8 (Decent Work and Economic Growth). The study highlights key research areas, including social entrepreneurship, innovation, impact investing, the plastics industry, the triple bottom line, and sustainable business models. This study clarifies the contributions of sustainable entrepreneurship to the SDGs, aiming to promote future advancements and informed decision-making in sustainable development initiatives. Rygh et al. (2021) argue that certain dominant paradigms currently restrict the impact of the international business field. The paper points out several ways international business (IB) research could better support the SDGs. Many SDGs, such as SDG 9 (Infrastructure, Industry, and Innovation), have international aspects, and IB's knowledge of global innovation and knowledge transfer is valuable. Second, they acknowledge that multinational enterprises (MNEs) have the power to both advance and impede SDG progress due to their vast resources and worldwide presence. The authors then propose recommendations to broaden the current IB perspective, such as: (1) adopting a more value-driven approach to the SDGs, (2) focusing more on the negative impacts of IB activities on sustainable development, (3) connecting micro and macro analyses through multi-level methods, and (4) examining IB's role in societal transformation and systemic change.

Celone et al. (2021) aim to explore how MNEs can better contribute to achieving the SDGs. They conclude that the most effective way to pursue the SDGs is through a comprehensive approach that acknowledges the significant influence and capabilities of MNEs. Progress toward the SDGs is mainly hindered by the complexity of the issues, a lack of genuine commitment to some goals in favor of profit, and limited long-term vision. However, clearly defining the problem and its features can help improve management strategies. Sinkovics et al. (2022) critically assess the role of international business in promoting the United Nations' SDGs. They emphasize the SDGs as a crucial framework for addressing global issues and stress the importance of businesses aligning their strategies with these goals to achieve real progress. Despite increased awareness, businesses often fall short in practice, leading to accusations of "SDG washing." Systemic challenges, including limited resources, a lack of integrated research, and inadequate institutional support, also hinder progress.

This study investigates how businesses can contribute to the SDGs through participation in Local Agenda 21 (LA21) initiatives, or equivalent local sustainability partnerships. Findings reveal that many SDG targets align with the primary reasons businesses engage in local sustainability partnerships. Furthermore, the outcomes achieved through participation in these partnerships enable businesses to further advance the SDGs. This research offers insights into how businesses can simultaneously achieve organizational goals and contribute to global sustainability, providing a model for fostering effective cross-sector collaborations that promote sustainable development (Ordonez-Ponce and Khare 2021).

3 Dimensions, Targets and Indicators of SDGs Related to International Business

Helble and Shepherd (2017) focus on the intricate connections between international trade and the achievement of SDGs related to poverty, hunger, inclusive growth, sustainable growth, education, health, and other linkages. The SDGs are divided into three dimensions of the Triple Bottom Line (TBL) framework: environmental, social, and economic dimensions, which include targets and indicators.

3.1 Environmental Dimension

Clean Water and Sanitation (SDG 6)

Ordonez-Ponce and Khare (2021) make a significant contribution to understanding the environmental implications of cryptocurrency trading, particularly Bitcoin, in the context of achieving SDG 6 (clean water and sanitation) and 13 (climate action). The empirical results demonstrate that Bitcoin trading activities are positively correlated with advancements in water and sanitation initiatives, yet simultaneously exert a detrimental impact on efforts to mitigate carbon emissions. These findings underscore the urgent need for the development of robust regulatory frameworks and the promotion of voluntary sustainable practices, particularly in emerging economies where institutional governance may be less well-developed. The study highlights the crucial role of policymakers, regulators, and investors in integrating sustainability principles into the rapidly evolving cryptocurrency landscape, ensuring that economic benefits do not compromise environmental integrity. Moreover, the research highlights the importance of integrating renewable energy solutions into cryptocurrency operations and promoting international collaboration to mitigate environmental impacts. Ultimately, while the economic potential of cryptocurrencies is considerable, achieving environmental sustainability in this sector remains a pressing challenge requiring coordinated, proactive strategies.

Abanyie et al. (2025) examined the interplay between water scarcity and sanitation outcomes in Damongo, Northern Ghana. The findings highlight profound deficiencies in water infrastructure and sanitation facilities, exacerbated by the privatization of water sources and irregular supply schedules. The reliance on pit latrines, coupled with limited household toilet access, significantly compromises public health, as evidenced by the high incidence of waterborne diseases such as typhoid and diarrhea. Extended water collection times further strain educational and economic activities, particularly for vulnerable groups. To address these challenges, the study advocates for strategic investments in sustainable water infrastructure, the implementation of equitable water pricing mechanisms with targeted subsidies, the prioritization of modern sanitation facility development, and the fostering of multi-stakeholder collaborations to promote resilient, health-oriented urban development.

Affordable and Clean Energy (SDG 7)

According to Sinha and Shahbaz (2018), a notable milestone in India's environmental performance is the threshold value of USD 2937.77 on the inverted U-shaped Environmental Kuznets Curve (EKC) for CO_2 emissions, signaling a critical transition point from national initiatives to greater engagement in international environmental efforts. The findings suggest that India is moving toward achieving sustainable, long-term prosperity. Renewable energy generation was found to have a significant and negative impact on emissions, both in the short and long term, while trade also contributed positively by reducing emissions. However, total factor productivity did not show a consistent effect. For policy, the study recommends a gradual shift from fossil fuels to renewable energy, rather than a rapid transition, to minimize the impact on economic growth. It proposes a system of discriminatory advances (loans) with subsidized interest rates, favoring rural households to support the adoption of renewable energy. Interest income from industry and urban areas could help fund rural transitions. Such phased policies could flatten the EKC, bringing the turnaround point closer within reach and enabling sustainable development without sacrificing economic progress.

Tamasiga et al. (2025) explore the relationship between SDGs and climate actions, particularly within the Just Energy Transition (JET), using a systematic review of existing literature. The research identifies significant synergies, such as the role of renewable energy in advancing both climate action (SDG 13) and access to clean energy (SDG 7). The transition to sustainable energy solutions is shown to support economic growth, poverty eradication, and the development of sustainable infrastructure. However, the review also highlights potential trade-offs, including the risk of increased inequality and economic disruption if the energy transition is not managed inclusively. The findings underscore the need for integrated policy frameworks that foster cross-sectoral collaboration, incentivize clean energy innovation, mobilize private capital, and implement targeted subsidies to ensure a just and equitable energy transition.

Climate Action (SDG 13)

Ulpiani et al. (2025) conducted a comprehensive analysis of the Expressions of Interest from 362 cities for the European Climate-Neutral and Smart Cities Mission, investigating the role of stakeholder engagement in urban climate action. The study employed descriptive statistics, network analysis using Gephi software, and text mining techniques to evaluate collaboration patterns. Key findings indicate that larger cities demonstrate more extensive collaboration frameworks. Citizens and national governments are identified as primary stakeholders; however, the engagement of entities such as utilities, energy communities, financial institutions, vulnerable groups, and trade unions is suboptimal. The research further reveals a nuanced landscape of stakeholder involvement, with academia, research institutions, and the private sector all making significant contributions to research and innovation activities. While intercity collaboration and networking are prevalent, the study underscores the imperative for enhanced integration of diverse stakeholders, particularly those representing marginalized populations. The results suggest a need for a transition toward more participatory engagement practices to effectively leverage the synergistic potential of inter-sectoral and inter-city collaborations in achieving urban climate neutrality.

Zhou et al. (2025) empirically examined the impact of China's Climate Resilient City Construction (CRCC) on corporate ESG performance using a difference-in-differences model with data from listed firms in China from 2013 to 2020. The study's findings indicate a significant positive relationship between the CRCC and corporate ESG performance, even after multiple robustness tests. Furthermore, the research identifies institutional investor shareholding and board diversity as key moderating factors that enhance the positive impact of the CRCC on corporate ESG performance. Mechanistic analysis reveals that the CRCC influences corporate ESG performance by increasing government financial incentives and promoting corporate green innovation. Heterogeneity analysis further demonstrates that the CRCC has a more pronounced effect on firms in regions characterized by severe meteorological hazards, high greenhouse gas concentrations, and specific climate zones, as well as firms with weak political connections and a high degree of digital transformation. The study concludes by offering insights for policymakers and market participants seeking to address climate change and foster sustainable development.

Life Below Water (SDG 14)

Moretti et al. (2024) developed a methodology to identify and prioritize the barriers hindering the progress of Sustainable Development Goal (SDG) 14 in Brazil. The methodology integrates the Fuzzy-DELPHI method, Borda's method, and Hesitant Fuzzy SWARA to prioritize barriers identified through a literature review and expert consultation. The results indicated that the most critical category of barriers is policy and management, and the most prominent barrier is the relaxation of the environmental licensing system for impactful activities. The research concludes that addressing the identified barriers is crucial to enhancing Brazil's progress toward SDG 14. Proposed solutions to mitigate these barriers include providing additional resources to environmental agencies, enhancing surveillance in protected areas, and implementing public policies to ensure regulatory compliance. The methodology developed in this research can be adapted and applied to other contexts, and future studies can assess the impact of the proposed actions and monitor Brazil's progress in achieving SDG 14.

Magris et al. (2019) developed a modeling approach to assess the impact of land-based mining catastrophes, specifically tailing dam failures, on marine ecosystems. The study integrated hydrological and coastal-circulation models to simulate the dispersal of pollutants, focusing on sediment transport, following a tailing dam spill event in the Doce River catchment in Southeastern Brazil. Key findings indicate that the pulse disturbance from the dam collapse led to a significant increase in sediment export, which impacted sensitive ecosystems, such as rhodolite beds. While the intensity of impacts diminishes over time, the total area of affected sensitive ecosystems is predicted to expand. The study concludes that the modeling approach effectively assesses the immediate and long-term impacts of mining-related sediment pollution on marine biodiversity. The results emphasize the need to develop restoration strategies and prioritize monitoring efforts, including within marine protected areas, to mitigate the adverse effects of mining activities and similar events. The methodology offers a valuable framework for quantifying the impacts of land-based disturbances on marine ecosystems and for optimizing recovery efforts.

Life on Land (SDG 15)

Geng et al. (2025) investigated the relationship between land use resilience (LUR) and land use efficiency (LUE) across 31 provinces in China. The study quantified LUR and LUE and analyzed their interconnectedness using the coupling coordination model. The research found that LUR in China increased from 0.2656 in 2000 to 0.3901 in 2022, and LUE rose from 0.7510 to 1.0845 during the same period. A positive correlation was observed between functional resistance and flow transformation in LUR and LUE. Furthermore, the LUR-LUE coupling coordination aligned more closely with the Sustainable Development Goals (SDGs) than models focusing solely on resilience or efficiency. The study concluded by categorizing LUR-LUE adaptive management optimization zones into six types: high-level synergy, high-level centrality, radiation-driven, radiation stagnation, radiation depression, and low-level solidification. These findings enhance the understanding of the dynamic nature of land use systems and support efforts toward high-quality, secure, and synergistic development.

Partnerships for the Goals (SDG 17)

Nielsen et al. (2023) conducted a systematic review of 132 articles published between 1991 and 2022 to analyze the intersection of disasters and international business (IB). The review distinguishes between man-made and natural disasters, identifying two main streams of research: one focusing on multinational enterprises (MNEs) and their strategic responses to disasters as exogenous shocks, and the other examining the broader role of MNEs and other actors in disaster contexts. The study highlights the increasing frequency and impact of disasters, such as the COVID-19 pandemic, on IB. The review reveals that while much of the research centers on MNEs' resilience and strategies in response to disasters, there's a growing recognition of the need to include a broader range of IB actors and to address the role of IB in disaster prevention, relief, and recovery. The authors conclude by emphasizing the importance of linking IB-disaster research with the United Nations Sustainable Development Goals (SDGs) to foster more holistic and impactful future research.

3.2 Social Dimension

No Poverty (SDG 1)

Brambilla and Porto (2017) analyze the multifaceted relationship between trade liberalization and poverty. It goes beyond simplistic narratives, acknowledging the potential for trade to lift people out of poverty through increased income opportunities and lower consumer prices, while also carefully considering the risks of job displacement and rising inequality, especially for vulnerable populations. The authors likely explore case studies and econometric evidence to illuminate the complex interplay of factors that determine the net impact of trade on poverty levels, examining factors like skill levels, access to credit, and institutional quality.

The SDGs aim to eradicate poverty in all its forms by 2030. The private sector plays a vital role in tackling global poverty. With increasing investment flows and expanding markets in developing countries, businesses are now among the most influential actors in shaping poverty outcomes. Companies can create positive change through their employment practices, access to essential goods and services, value chain operations, and engagement in public policy. Yet, most businesses still lack a deep understanding of their actual impact on poverty.

The Poverty Footprint tool was developed as a collaboration between the UN Global Compact and its partners to help companies, alongside civil society organizations, evaluate how their operations and value chains affect people living in poverty. This approach looks beyond income alone to examine the multifaceted nature of poverty. It enables companies to assess their impacts in areas such as livelihoods, empowerment, health, stability, and equality, and it helps generate data-driven recommendations to reduce harm and improve outcomes. The Poverty Footprint process produces a comprehensive understanding of how a company and its value chain impact people living in poverty, both positively and negatively. Through this collaborative research, companies and their civil society partners identify key actions to enhance beneficial effects and mitigate harm. The insights can guide improvements in business strategies and models that align with development goals while supporting business growth. The outcome is a jointly authored public report that outlines findings, commitments, and follow-up actions, thereby fostering greater transparency, accountability, and trust with affected communities. By engaging local stakeholders and conducting in-depth analysis, the process fosters mutual learning. It enables the development of inclusive business approaches that can make a meaningful contribution to poverty eradication.

Zero Hunger (SDG 2)

The report, commissioned by the World Business Council for Sustainable Development (2024) and conducted by KPMG, examines the complex web of risks facing the food retail sector as it seeks to integrate nature-positive practices into business models. The article highlights that the food retail sector faces increasingly interconnected risks related to nature, such as climate change, biodiversity loss, and ecosystem degradation, which threaten long-term business viability. Traditional business models, focused on short-term profits, are unsustainable and undermine both environmental and economic resilience. The report emphasizes that climate and nature risks are deeply linked, requiring holistic, long-term strategies and stronger partnerships across the supply chain, especially with farmers. It also stresses the need for new metrics to measure success, greater internal awareness of environmental impacts, and collaborative efforts throughout the value chain to drive systemic change toward sustainability. Therefore, this analysis recommends that food retailers shift from short-term, cost-focused procurement to building long-term, sustainable relationships with suppliers. It urges companies to integrate nature-related risks into their overall risk management, invest in partnerships and technology to support sustainable farming practices, and raise internal awareness of environmental issues across all departments. Collaboration throughout the value chain is crucial for integrating nature-positive practices and mitigating interconnected risks, thereby promoting systemic change and fostering long-term business resilience.

Kumar et al. (2023) explored the enablers of food security, safety, and quality that transform the green food supply chain management (GFSCM) through the Internet of Things (IoT) and blockchain technologies to reduce hunger. The study identified twelve enablers and used a combined Interpretive Structural Modeling (ISM) and Analytical Network Process (ANP) technique to examine their interrelationships and significance. The findings reveal that IoT and blockchain technologies are the main drivers of the contemporary GFSCM system. The research concludes that the implementation of GFSCM practices helps reduce food wastage and carbon emissions, while also improving food quality and safety. The study emphasizes the importance of digital transformation in achieving food security, safety, and quality, and suggests management, theoretical, and policy implications toward achieving zero hunger.

Good Health and Well-Being (SDG 3)

World Business Council for Sustainable Development (2024) notes that due to the complexity of food systems, no single entity can tackle the challenges alone. Food retailers and suppliers, however, are strategically positioned to connect producers and consumers, enabling them to drive impactful changes within supply chains through individual and collaborative efforts. To promote environmental sustainability, companies must strengthen their relationships with farmers by offering both financial and non-financial incentives for adopting sustainable farming practices and investing in technology to enhance supply chain transparency and outcomes.

Duarte et al. (2023) assessed the relationship between consumer vulnerability (CV) and well-being (WB) by comparing ordinary (non-pandemic) and pandemic consumption contexts among Portuguese and Brazilian consumers. The study analyzed data from a cross-cultural sample of 397 consumers using structural equations modeling. The results revealed an inverse relationship between CV and well-being, which worsened during the pandemic. The research concluded that Refund Policies, Product Promotions, and Purchase Ability were the CV dimensions most affected by the COVID-19 outbreak. Furthermore, fear was found to partially mediate the effect of vulnerability on well-being. The findings offer valuable insights for companies and public institutions in planning effective responses to future disruptive events.

Quality Education (SDG 4)

Larsen (2025) investigated the implementation of Project-Based Learning (PjBL) in business and management education using a scoping review methodology. The review synthesized findings from 51 empirical and non-empirical studies, identifying key themes including publication trends, theoretical foundations, and design and implementation features of PjBL. The analysis highlighted a wide variance in how PjBL is conceptualized and applied, revealing the necessity of further theoretical development to define its distinct role within business and management education. The research concluded that PjBL offers notable benefits, such as enhanced student engagement and competency development. However, significant challenges persist, including resource intensity, project complexity, and assessment issues. The review consolidated existing knowledge and proposed a structured research agenda to guide future studies and improve PjBL research and practice.

Narong (2025) conducted a review of business and management education for sustainability, analyzing 365 Scopus-indexed documents. The study used descriptive statistics, citation analysis, co-citation analysis, thematic analysis, and science mapping. The results of the analysis reveal that the body of knowledge remains underexplored and is concentrated in economically developed regions. The intellectual structure of the field is built around three schools of thought: Responsible Management Education, Holistic Management Education, and Higher Education for Sustainable Development. The study concludes that there is a growing number of studies in the field. Major topics of interest include the SDGs, higher education, corporate social responsibility, and principles for sustainable management education. The author emphasizes the need for research from less developed regions, increased international collaboration, and a greater focus on emerging areas, such as supply chain management, climate change, and the impacts of COVID-19.

Gender Equality (SDG 5)

Shepherd and Stone (2017) shed light on the gender dimensions of trade, examining how trade policies and practices can either empower women economically or exacerbate existing gender inequalities. It likely explores women's diverse roles in the global trading system—as producers, consumers, and entrepreneurs—examining the barriers they face in accessing markets, credit, and technology. The authors might analyze how trade can create opportunities for women-owned businesses and contribute to greater gender equality in the labor market, while also addressing potential adverse impacts, such as exploitation and precarious employment.

Donnelly et al. (2025) explored gender equality within transnational Professional Services Firms (PSFs) through an integrative literature review. The review examines how gendered structural and relational processes and practices at multiple levels shape the experiences and behaviors of professionals in these firms. The investigation explains the shortcomings of current gender equality research and strategy. The authors developed a new framework for the transnational analysis of gender equality, synthesizing the dynamic processes and practices that create gendered PSFs. The study concludes by setting out three prospective agendas for the international business community to advance future gender equality research, strategy, and practice.

Ilie and Mondragón Quintana (2023) found that the EMBA program positively influenced participants' thinking styles, fostering a more integrative thinking model beneficial for top executive roles. Notably, the program enhanced women's strategic and risk-taking thinking. However, the curriculum design inadvertently widened the gender gap in analytical skills, favoring men, and reinforced women's pre-existing strengths in execution and relationship development. Conversely, the EMBA program did not adequately develop the organizational and social thinking preferences of male students. This disparity contradicts the Principles for Responsible Management Education (PRME) agenda, highlighting a failure to fully prepare both genders for top management positions. The authors concluded that management education programs should intentionally design curricula to balance the development of diverse thinking styles, thereby promoting gender equality.

Gannon et al. (2025) explored the influence of female representation in micro and small business leadership on climate change adaptation in Senegal and Kenya. The study revealed that businesses with female leaders tend to adopt more sustainable and fewer unsustainable adaptation strategies when faced with more frequent extreme weather events. This suggests that female leadership may enhance adaptive capacity and long-term resilience. The research also indicated that adaptation assistance, such as training and support, can further promote the adoption of both sustainable and unsustainable strategies, particularly in businesses with female leaders. The authors concluded that promoting an inclusive business environment that supports women entrepreneurs is crucial for advancing climate justice and improving resilience. These findings underscore the importance of incorporating gender dynamics into climate change adaptation strategies within the private sector. However, the study acknowledges limitations, including the sample size and geographical scope, suggesting a need for further research to validate these findings across diverse contexts.

Sustainable Cities and Communities (SDG 11)

Wang et al. (2024) explored the roles of various actors in a community-based waste management experiment in Qingdao, China. The study revealed that NGOs, the city government, industries, and alliances among them played critical roles in accumulating resources, enhancing networks, sustaining coordination, and creating long-term linkages. These actors significantly shaped the development and outcomes of the experiment, with their roles and relative importance evolving throughout its progression. The authors concluded that while effective solutions need to be context-specific, the study exemplifies a pathway for transforming community waste management in Chinese cities, which face challenges such as top-down governance and minimal public participation. The research highlights the importance of understanding the evolving roles of actors in promoting the success of sustainability experiments, providing valuable insights for policymakers and practitioners.

Antolín-López et al. (2024) examined the relationships between urban sustainability and city livability in European cities, taking into account the moderating role of business regulatory quality. The study revealed that social urban sustainability has a positive predictive value for city livability over time. However, the influence of economic and environmental urban sustainability on city livability is more complex and contingent on business regulatory quality. Specifically, economic and environmental urban sustainability have a more substantial positive effect on city livability when business regulatory quality is high. Conversely, the study found that low business regulatory quality weakens the positive impact of economic urban sustainability and can even lead to an adverse effect on environmental urban sustainability, ultimately affecting city livability. The authors concluded that a practical and supportive regulatory framework is crucial for translating economic and environmental sustainability efforts into tangible improvements in residents' well-being.

Peace, Justice and Strong Institutions (SDG 16)

Dave et al. (2024) examined the peacemaking and peacebuilding endeavors of two American businessmen, Elon Musk and Armand Hammer, respectively, highlighting the constraints imposed by international relations (IR). The analysis of the cases revealed that both businessmen faced limitations in achieving their peace objectives due to the realities of IR, specifically international law in Musk's case and international society in Hammer's. The study further suggests that effective peace initiatives by business leaders necessitate transforming leadership characterized by moral imagination. The authors concluded that while business leaders can play a role in peace initiatives, their efforts are often more fruitful in peacebuilding, which involves fostering conditions for peace, rather than peacemaking, which entails direct conflict resolution. This observation underscores the complex interplay between business, peace, and the inherent challenges posed by the frontiers of IR.

Ramirez et al. (2022) conducted a qualitative study from 2019 to 2022, analyzing Colombia's energy policy within the context of sustainable development and renewable energy diversification. The study's findings indicate that weak institutions pose a significant impediment to Colombia's energy transition. It also highlighted that the Western concept of energy democracy, which presumes effective governance aligned with SDG 16 (Peace, Justice, and Strong Institutions), is challenged by social unrest, violence, corruption, and inequality in Colombia. The researchers concluded that

Colombia's energy policy views green investment as a market opportunity rather than a tool for institutional building or phasing out fossil fuels, and it lacks meaningful engagement with civil society, particularly indigenous peoples, in developing small-scale green initiatives. They argue that to achieve peaceful energy diversification, Colombian energy policy needs to address SDG 16 more directly by focusing on peace transitions and building robust institutions to support inclusive societies.

Partnerships for the Goals (SDG 17)

Van Dorp et al. (2025) address the challenge faced by companies and investors in defining fundamental concepts and methodologies for operating in fragile and conflict-affected settings, given the proliferation of competing guidelines and standards for responsible business behavior and social impact management. The authors outline an approach that utilizes forward-looking human security partnerships between businesses and local stakeholders to identify and assess the potential peace value and risks of business interventions as they emerge over time. The study concludes that businesses can utilize novel governance arrangements to design and measure social impacts that foster peace through improvements in human security. It argues that human security partnerships, which entail long-term dialogue between businesses and conflict-affected communities, are a means to bridge high-level policy and management goals, facilitating meaningful transformation at the community level.

3.3 Economic Dimension

Decent Work and Economic Growth (SDG 8)

Vandenberg (2017) addresses the critical question of whether trade can result in net employment gains. It moves beyond simple assertions about job creation, carefully weighing the job creation and job destruction effects of trade liberalization. The author will probably explore the types of jobs created and lost, the skills required for new opportunities, and the role of complementary policies (such as education and training programs, labor market reforms, and social safety nets) in maximizing the employment benefits of trade and mitigating its negative consequences.

The United Nations Global Compact (2025) initiative encourages businesses to ensure all employees earn a living wage by 2030. A living wage is defined as sufficient remuneration for a standard workweek to afford a decent standard of living, covering essential needs like food, housing, healthcare, and education. The key components of achieving the Living Wage Ambition include defining a living wage as sufficient remuneration for a decent standard of living, recognizing its importance in advancing Sustainable Development Goals, focusing initially on direct employees and contractors while extending to the entire supply chain, developing fair compensation policies with stakeholder engagement, assessing the gap between current wages and living wages using reliable benchmarks, and understanding challenges like the lack of a universally agreed-upon living wage level. Companies such as Inditex, L'Oréal, Schneider Electric, and Unilever are already taking action through collective bargaining, responsible purchasing practices, and strategic partnerships.

The Global Wage Report 2024–2025 analyzes recent wage trends and labor income inequality. It finds that global real wage growth recovered in 2023, with a projected increase in 2024, though disparities persist between advanced and emerging G20 economies. Wage inequality remains high, with low-income countries displaying the highest levels. While most countries adjusted minimum wages in response to inflation, these changes often didn't fully compensate low-wage earners. The report also reveals that labor productivity in high-income countries has increased more rapidly than real wages over the long term. Notably, wage inequality has declined in many countries since the beginning of the twenty-first century, particularly in low-income nations. However, the global wage distribution still shows significant inequality, with a large proportion of workers concentrated at the low end. Including non-wage workers in the analysis further increases labor income inequality (International Labour Office 2025).

Industry, Innovation, and Infrastructure (SDG 9)

The analysis of case studies (OECD 2024) in this toolkit reveals that successful industry decarbonization relies on a combination of key elements. Knowledge-sharing platforms and matchmaking initiatives facilitate the efficient connection of stakeholders and resources, while multi-stakeholder collaboration among governments, private actors, and financial institutions enhances implementation. Public support, especially when targeted and time-limited, is essential for unlocking private investment, and de-risking instruments are particularly effective in optimizing the use of public funds. Additionally, building institutional and industry capacity, developing sustainable taxonomies, and using innovative financial tools—especially for early-stage technologies—are vital for accelerating progress. Ultimately, establishing robust pipelines of viable projects is essential to facilitate the replication and scaling of decarbonization solutions across various sectors and regions. The analysis of case studies in this toolkit reveals that successful industry decarbonization relies on a combination of key elements. Knowledge-sharing platforms and matchmaking initiatives facilitate the efficient connection of stakeholders and resources, while multi-stakeholder collaboration among governments, private actors, and financial institutions enhances implementation. Public support, especially when targeted and time-limited, is essential for unlocking private investment, and de-risking instruments are particularly effective in optimizing the use of public funds. Additionally, building institutional and industry capacity, developing sustainable taxonomies, and using innovative financial tools—especially for early-stage technologies—are vital for accelerating progress. Ultimately, establishing robust pipelines of viable projects is essential to facilitate the replication and scaling of decarbonization solutions across various sectors and regions.

Rukanova et al. (2023) explored the concept of public value creation through voluntary information sharing between businesses and government entities, facilitated by digital infrastructure innovations. The study's results emphasize a shift from the traditional view of public value creation, which primarily focused on government and public services, toward a more interactive perspective. The findings highlight that public value creation should be understood as a collaborative and dynamic process involving active participation and engagement from both government and business actors. The research effectively demonstrates how voluntary information sharing, enabled and supported by digital infrastructures, can significantly contribute to the creation of public value in various contexts. In their concluding discussion, the authors emphasize the importance of recognizing and embracing public value creation as an inherently interactive process that requires close collaboration and information exchange between government and business stakeholders. They posit that the framework developed within the study provides a valuable tool for analyzing and understanding the dynamics of public value creation in the context of voluntary business-to-government information sharing, offering insights for both researchers and practitioners.

Reduced Inequalities (SDG 10)

Urata and Narjoko (2017) offer a detailed analysis of the complex relationship between trade and income inequality, both within and between countries. It examines the factors that contribute to the unequal distribution of trade gains, including differences in skills, access to capital, and market power. The authors likely examine how trade can impact the wage gap between skilled and unskilled workers, the concentration of wealth, and the competitiveness of different industries. The chapter may also explore the role of trade policies and domestic institutions in shaping the distributional outcomes of trade liberalization.

Johannesen (2025) explored the application of feminist participatory action research (FPAR) within the International Council of the Exploration of the Sea (ICES) to advance gender equality in international ocean science. The study's results revealed that the stages of awareness, understanding, and action in advancing a gender equality agenda are dynamic and overlapping, indicating that change processes are not linear but rather iterative and complex. Despite encountering varying degrees of short-term acceptance and resistance, the research emphasized the critical importance of ongoing community engagement, monitoring, and evaluation for fostering longer-term and sustainable change within the organization. The analysis demonstrated how FPAR can be effectively applied within an intergovernmental organization, highlighting the contribution of multiple factors, including drivers for change, systematic data collection, collaborative co-creation through dialogue, and the implementation of specific actions, in driving the development and publication of a gender equality plan for the organization. In discussing the broader implications of these findings, the author concluded that meaningful actions to promote gender equality can occur at various scales, with participatory engagement serving as a central and essential component. The study's results provide empirical evidence on how gender equality, a globally recognized meta-norm, is translated into organizational practice. However, the author also critically discussed the limitations of formal gender equality plans in achieving substantive and lasting change, pointing to the need for a more nuanced and context-sensitive approach to address the complex and persistent challenges of gender inequality within international marine science organizations.

Responsible Consumption and Production (SDG 12)

Understanding and managing environmental risks is fundamental to ensuring long-term business viability. With more than half of the world's total GDP moderately or highly dependent on nature and its services, the economy is highly exposed to risks from the unprecedented loss of nature, ecosystems, and biodiversity in recent years. The current and expected impacts of nature on the resilience of corporate business models and their long-term value creation are becoming increasingly complex, making it difficult for any business to ignore, particularly given their interconnected nature. They can impact different operational functions, supply chain entities, geographies, and social groups in various ways. In particular, the connection between climate and nature risk cannot be underestimated. Degradation of ecosystems poses substantial risks to the food sector, which heavily depends on nature and ecosystem services to function. By proactively identifying and addressing climate and nature-related risks, businesses within the food system can help mitigate the most significant global risks in the coming decade (World Business Council for Sustainable Development 2024).

Shaikh et al. (2024) addressed the critical intersection of SDG 12, which calls for responsible consumption and production, and the environmental limits of natural resource use, particularly cropland within the global agri-food system. The study employed a multi-regional input–output model to examine how virtual cropland flows, embedded in agri-food commodities, impact national consumption and production-based environmental limits. A key finding was that certain countries contribute to the exceedance of producer countries' production-based cropland limits, regardless of whether the consumer countries exceeded their consumption-based limits. For instance, 86% of China's consumption-based cropland footprint originated from regions exceeding their production-based cropland limits, despite China's per capita consumption footprint being within safe limits. This was attributed to the consumption of commodities like oilseeds, cereals, and fruits and vegetables, both domestically produced and imported. The research also identified international trade patterns that can inform interventions for responsible agri-food consumption and production at the national level, aligning with SDG 12. In conclusion, the authors emphasized the need for more stringent regulations and improved policies to mitigate the exceedance of both consumption-based and production-based environmental limits. This is essential to avoid breaching the global land-system change planetary boundary. The study highlighted the interconnectedness of the agri-food system, where production, consumption, and trade have substantial environmental impacts on land, water, and other natural resources. By assessing cropland use against national cropland limits, the research provides policymakers with insights to develop effective measures, such as regulations, taxes, subsidies, and incentives that promote responsible agri-food consumption and production.

Partnerships for the Goals (SDG 17)

Moharrak et al. (2025) explored the emergence of international hidden entrepreneurs (IHEs) in Saudi Arabia through the lens of institutional theory. The study defines IHEs as international entrepreneurs who form concealed partnerships with local entrepreneurs to circumvent stringent local regulations that impose high costs on foreign business owners. The research highlights that IHEs primarily originate from emerging market nations and thrive in environments with complex institutional structures. The findings reveal that IHEs emerge as a response to regulatory barriers, socio-cultural norms, and cognitive skill gaps. While IHEs contribute to the local economy by providing expertise and services, their activities also present ethical and economic challenges, including hindering the development of local talent and creating unfair competition. The study concludes by emphasizing the need for policymakers to address the complex dynamics of IHEs, striking a balance between economic aspirations and the ethical development of the local market.

4 Specific Indicators for Economic Sector (Scope and Subject Wise)

Agricultural Sector

Martin (2019) examines the pivotal role of agricultural trade in bolstering food security and combating hunger. It likely investigates the influence of trade policies (e.g., tariffs, subsidies) on agricultural markets, as well as the effects of price volatility and supply chain inefficiencies on the availability, affordability, and stability of food supplies, particularly in developing countries. Likewise, the research examines the implications of trade for both food-exporting and food-importing nations, taking into account the specific challenges faced by smallholder farmers and net food-importing developing countries.

Industry Sector

The Garment and Footwear Sector

OECD (2024) outlines essential points for brands and retailers to consider when using sustainability certifications as part of their due diligence efforts. It also provides guidance for policymakers. The garment and footwear industry makes a significant contribution to economic and social progress in developing countries, particularly through job creation and skill development. However, the sector is also confronted with serious challenges. Its vast, complex, and often fragmented supply chains, coupled with economic pressures and unstable markets, make it difficult for companies to correctly identify and respond to risks related to labor, human rights, the environment, and business integrity. The legal context and enforcement mechanisms in sourcing countries also shape how companies approach due diligence. To manage these complex challenges and communicate responsible practices, many brands and retailers have increasingly relied on third-party sustainability certifications. According to a survey conducted for this paper, 78% of brands and retailers found certifications helpful in supporting human rights and environmental due diligence. Nonetheless, concerns persist among policymakers, civil society, and other stakeholders regarding the quality, credibility, and appropriate role of certifications. Reports have highlighted issues like weak assessment procedures, conflicts of interest, and over-reliance on certifications by companies. While certifications can be helpful, they should not be treated as a complete solution or a "safe harbor." International standards emphasize that companies must not rely solely on certifications, even those with complex supply chains. Instead, they should apply a risk-based approach and prioritize the most significant risks. This includes conducting independent assessments, engaging stakeholders meaningfully, using grievance mechanisms, and maintaining open communication with suppliers.

Manufacturing Sector

Bonfanti et al. (2023) explore how manufacturing companies operationalize sustainable business models (SBMs) to contribute to achieving the United Nations' SDGs. It identifies a diverse range of sustainable practices implemented by these companies, revealing a commitment not only to economic and environmental sustainability but also to social dimensions, such as enhancing employee well-being, improving workplace conditions, and fostering engagement. The research highlights the adoption of local embeddedness strategies, networking development, and the creation of sustainable ecosystems, extending the triple bottom line framework (economic, environmental, and social dimensions). Manufacturing companies were found to contribute to 11 of the 17 SDGs, demonstrating areas of substantial impact and opportunities for further enhancement. The study emphasizes the importance of integrating sustainability into core business strategies to make meaningful contributions toward global development goals, as outlined by the dimensions of the triple bottom line (TBL) framework.

Giacosa et al. (2024) explored the intersection of additive manufacturing (AM) and international business, using thematic and content analysis of academic and professional sources to bridge the gap between theory and practice. The study reveals that AM offers benefits such as shortening the value chain, enhancing customer relationships through customization, and promoting more sustainable production. However, it also identifies challenges that may hinder the adoption of AM in international business. The authors conclude that AM has the potential to be disruptive in international business, influencing product specialization, production dispersion, and training needs. The research emphasizes the importance of an intensified industry-academia dialogue to support AM adoption and develop a comprehensive understanding of its implications for the global strategy of international companies.

Public and Services Sector

Ehigiamusoe et al. (2025) explored the intricate relationship between service sector development, government policies, and environmental quality in Southeast Asia. The study's findings suggest that while the expansion of the service sector can stimulate economic growth, it also contributes to environmental degradation through increased carbon emissions and a larger ecological footprint. Specifically, sub-sectors like financial services, tourism, transport, ICT, and trade were identified as significant contributors to carbon emissions. This environmental impact is primarily attributed to the service sector's heavy reliance on energy consumption, predominantly from fossil fuels, to power energy-intensive equipment and support activities such as transportation and waste management. Despite the expectation that a shift toward a service-oriented economy might reduce emissions, the study reveals that the service sector's growth in Southeast Asia remains heavily dependent on fossil fuels, thereby exacerbating environmental concerns. In discussing policy implications, the authors emphasize the need for a balanced approach that acknowledges both the economic and environmental imperatives of the service sector. They suggest that policymakers should implement incentives to encourage environmentally friendly practices within the service industry, such as adopting clean and energy-efficient technologies and transitioning to low-carbon transport. Furthermore, the study emphasizes the significance of government consumption expenditure and tax revenue as tools to foster the growth of the service sector while also mitigating its environmental impact. The researchers advocate for a long-term, comprehensive emissions-reduction strategy that includes investments in clean production and renewable energy, coupled with stringent environmental regulations and public awareness campaigns.

Iqbal et al. (2025) explored the development of a sustainable governance model for Municipal Solid Waste (MSW) management in Pakistan, aligning the sector with the country's evolving economic goals. The study leveraged discrete mathematics and logical frameworks to demonstrate the potential of the waste sector in reducing public expenditure and fostering economic sustainability. The proposed Wasteism Hierarchy Approach (WHA) provides a structured set of policy guidelines spanning various waste hierarchy levels, designed to support informed decision-making for municipalities, waste management companies, and local governments. Key considerations within this approach include the interplay of waste collection modes, fleet management strategies, infrastructure development, and the delineation of waste catchment areas. Furthermore, the research incorporated a Multi-Criteria Decision-Making (MCDM) framework to rank and prioritize feasible MSW treatment and disposal options, ensuring a systematic and transparent evaluation process. The authors concluded that integrating the waste sector into the circular economy presents a significant opportunity to address climate change, promote regenerative agriculture practices, and support local economic sustainability in Pakistan. The research emphasizes the transformative potential of aligning MSW management goals with both international climate commitments and national policies. Effective governance models are crucial for achieving this alignment, requiring a holistic approach that integrates organizational capacity, environmental mitigation strategies, cost-effective solutions, and active community involvement throughout the planning and execution phases. The study advocates for a multilayered framework that considers technology, organization, and environment to identify and address gaps and risks that may impede the development of sustainable and smart city waste governance.

Fishery Sector

Jolaosho et al. (2025) synthesized the existing literature on microplastics (MPs) in both freshwater and marine ecosystems across all six continents, highlighting the interconnectedness of these environments in terms of MP contamination. The review revealed that the most commonly produced plastic polymers, polyethylene (36%) and polypropylene (21%), are also the most abundant MPs found in aquatic ecosystems. The study emphasized the diverse sources of primary and secondary MPs, including land-based disposal, ocean sources, airborne deposition, wastewater treatment facilities, automobiles, pharmaceuticals and personal care products, and synthetic textiles. The findings highlight the pervasive nature of MP pollution and the numerous pathways through which these pollutants enter and accumulate in aquatic environments, posing ecological risks to various species. In discussing the implications of these findings, the authors stress the urgency of addressing the microplastic crisis through effective mitigation strategies and policy interventions. They highlight the importance of understanding the distribution dynamics, fate, and transport processes of MPs to develop targeted solutions. The review emphasizes the importance of collaborative efforts across various sectors and levels of governance, from local to global, to implement sustainable practices and policies that reduce plastic waste and mitigate the environmental impacts of microplastic pollution.

Coronado et al. (2024) employ an institutional mapping approach to analyze the Mexican governmental system's response to the coexistence of fisheries and oil industry operations in the southwestern Gulf of Mexico, focusing on the implementation of the SDGs. The research identifies five sectors directly responsible for regulating these industries, encompassing 37 institutions and agencies. The findings reveal that the Mexican government places greater emphasis on economic development and infrastructure (SDGs 8 and 9). At the same time, comparatively less attention is given to food security and gender equality (SDGs 2 and 5). This prioritization has led to an imbalance in decision-making processes, with economic objectives often taking precedence over social and environmental considerations. The study highlights the complexity of the institutional network governing the fisheries-oil system. It underscores the challenges associated with achieving sustainable development in regions where extractive industries and traditional livelihoods intersect. Institutional performance is critical for regional sustainable development, especially in areas rich in both biodiversity and hydrocarbon resources. The research demonstrates that, despite efforts to align national policies with the 2030 Agenda, including reforms to planning laws and the establishment of intersectoral commissions, substantial challenges remain. Ineffective institutional networks, often marked by limited transparency and weak enforcement, have led to resource degradation and increased social vulnerability for local communities. The study calls for a more balanced approach that strengthens the roles of institutions responsible for social and environmental outcomes, ensuring that the benefits of economic growth do not come at the expense of ecosystem health and community well-being.

Transport Sector

Naseri et al. (2025) develop a comprehensive model to predict household transportation life-cycle greenhouse gas (GHG) emissions, with a focus on the city of Montreal, Canada. By leveraging a large-scale origin–destination survey and integrating additional datasets on built and social environments, the research evaluates the GHG emissions generated by all possible transportation modes used by households, including personal vehicles, buses, subways, bicycles, and diesel trains. Unlike most previous studies, which primarily concentrate on private vehicle emissions or only consider tailpipe emissions, this work calculates full life-cycle emissions, revealing that relying solely on tailpipe data underestimates total GHG emissions by ~20%. The study further distinguishes itself by applying and comparing ten advanced machine learning algorithms, including Light Gradient Boosting Machine and Categorical Boosting, to determine the most accurate model for predicting household transportation emissions. To interpret these complex models, Shapley Additive exPlanation (SHAP) values and Partial Dependence Plots (PDP) are used, identifying and illustrating the relative influence and direction of key variables affecting household GHG emissions. The findings underscore the significant impact that a shift from gasoline vehicles to electric vehicles could have, estimating a 57% reduction in transportation GHG emissions if such a transition were fully realized in Montreal. The research also explores the modal shifts necessary to achieve the Canadian government's transportation GHG reduction targets, providing scenario analyses that inform policy and planning. By considering a wide range of variables, many of which have been overlooked in prior research, this study offers a nuanced understanding of the determinants of household transportation emissions. It highlights the importance of adopting life-cycle perspectives and advanced machine learning techniques in environmental impact assessments. The results underscore the need for comprehensive strategies that extend beyond vehicle electrification, encompassing broader changes in travel behavior and urban infrastructure to achieve ambitious climate goals.

Tourism

According OECD (2024), tourism is a cornerstone of Spain's economy and a key driver of regional development, but Spanish regions face challenges in monitoring the effectiveness of tourism policies and measuring the sector's economic, social, and environmental impacts. This report introduces a system of 30 indicators and 57 metrics, informed by key policy priorities identified in existing tourism strategies and measurement frameworks. The indicators cover economic benefits, seasonality, local resident perceptions, accessibility, climate change mitigation, and water management. A subset of ten priority metrics is highlighted to monitor critical policy issues and guide action toward sustainable development. The framework is built on international best practices and tailored to the unique characteristics of the participating regions, thereby facilitating both local relevance and international comparability. The report includes detailed guidance for compiling core and supplementary indicators. It outlines future development opportunities to refine methodologies and address data gaps, gaps-particularly in areas such as cultural heritage, greenhouse gas emissions, waste, and digitalization. Ultimately, the framework aims to support evidence-based policymaking and promote more balanced, sustainable tourism development models in Spain.

Closing Remarks

The practical implementation of the SDGs requires a radical transformation in business operations, one that transcends short-termism and integrates sustainability as an irreducible strategic axis. Companies, key catalysts, must adopt triple-impact models—encompassing economic, social, and environmental aspects—linking innovation with equity, as demonstrated by advances in clean energy, regenerative agriculture, and circular waste management. However, overcoming "SDG washing" requires rigorous audits, genuine partnerships with communities and governments, and internal policies that prioritize living wages, diversity, and ecosystem restoration. Success will depend on institutionalizing ethics in global value chains, where technology and inclusive governance amplify positive impacts. Only then will the promise of the 2030 Agenda be realized: resilient development that leaves no one behind and respects planetary boundaries. The time to act, with urgency and coherence, is now.

References

Abanyie SK, Amuah EEY, Nang DB (2025) Water scarcity and its implications on sanitation: a perspective study in an emerging city in Northern Ghana. Green Technol Sustain 3(2):100138. https://doi.org/10.1016/j.grets.2024.100138

Antolín-López R, Martínez-Bravo MDM, Ramírez-Franco JA (2024) How to make our cities more livable? Longitudinal interactions among urban sustainability, business regulatory quality, and city livability. Cities 154:105358. https://doi.org/10.1016/j.cities.2024.105358

Arana C, Franco IB, Joshi A, Sedhai J (2020) SDG 15 life on land. In: Franco IB, Chatterji T, Derbyshire E, Tracey J (eds) Actioning the global goals for local impact: towards sustainability science, policy, education and practice. Springer, Singapore, pp 247–264

Arte P, Wang Y, Dowie C, Elo M, Laasonen S (2023) Sustainability in international business: an introduction. In: Arte P, Wang Y, Dowie C, Elo M, Laasonen S (eds) Sustainable international business: smart strategies for business and society. Springer, Berlin, pp 1–11

Bonfanti A, Mion G, Brunetti F, Vargas-Sánchez A (2023) The contribution of manufacturing companies to the achievement of sustainable development goals: an empirical analysis of the operationalization of sustainable business models. Bus Strat Environ 32(4):2490–2508. https://doi.org/10.1002/bse.3260

Brambilla I, Porto G (2017) Trade and poverty reduction. In: Win–win: how international trade can help meet the sustainable development goals, pp 61–87

Celone A, Cammarano A, Caputo M, Michelino F (2021) Is it possible to improve the international business action towards the sustainable development goals? Crit Perspect Int Bus 18(4):488–517. https://doi.org/10.1108/cpoib-08-2020-0122

Coronado E, Zepeda-Domínguez JA, Espinoza-Tenorio A, Santamaria DC, Ramos-Muñoz D, Monzón-Alvarado C (2024) Institutional mapping and implementation of the sustainable development goals across co-existing industries: the case of the fisheries-oil system in Mexico. Extract Ind Soc 17:101390. https://doi.org/10.1016/j.exis.2023.101390

Dave A, Simha A, Topuzova L (2024) Pursuit of peace by business leaders: frontiers of international relations. Bus Horiz 67(6):769–782. https://doi.org/10.1016/j.bushor.2024.08.006

Donnelly R, Hebson G, Chaudry S (2025) Transnational gender equality: a framework for analysis and three prospective agendas for future IB research. Int Bus Rev 34(3):102417. https://doi.org/10.1016/j.ibusrev.2025.102417

Duarte P, Linardi MA, Domingues HS, Silva SC (2023) Consumer vulnerability and well-being across contexts: implications for international businesses. Heliyon 9(3):e14612. https://doi.org/10.1016/j.heliyon.2023.e14612

Ehigiamusoe KU, Lee C-C, Lean HH (2025) Analysis of the economic and environmental imperatives of the service sector: the role of government in promoting sustainable development. J Environ Manag 376:124470. https://doi.org/10.1016/j.jenvman.2025.124470

Gannon KE, Eskander SMSU, Avila-Uribe A, Castellano E, Diop M, Agol D (2025) The role of gender in firm-level climate change adaptation behaviour: insights from small businesses in Senegal and Kenya. Clim Risk Manag 48:100699. https://doi.org/10.1016/j.crm.2025.100699

Geng Y, Li X, Chen J (2025) Integration of land use resilience and efficiency in China: analysis of spatial patterns, differential impacts on SDGs, and adaptive management strategies. Appl Geogr 175:103490. https://doi.org/10.1016/j.apgeog.2024.103490

Giacosa E, Crocco E, Kubálek J, Culasso F (2024) Additive manufacturing in international business: Bridging academic and practitioners' perspectives. J Int Manag 30(3):101151. https://doi.org/10.1016/j.intman.2024.101151

Helble M, Shepherd B (2017) Win–win: how international trade can help meet the sustainable development goals. Asian Development Bank Institute. https://www.adb.org/sites/default/files/publication/327451/adbi-win-win-how-international-trade-can-help-meet-sdgs.pdf

Huaman-Pastorelli S (2020) Environmental literacy of Peruvian middle school graders and their teachers of private and public schools. Harvard University. https://nrs.harvard.edu/URN-3:HUL.INSTREPOS:37364880

Ilie C, Mondragón-Quintana JC (2023) Is sustainable management education contributing to gender equality? Int J Manag Educ 21(3):100886. https://doi.org/10.1016/j.ijme.2023.100886

International Labour Office (2025) Global wage report: executive summary. https://www.ilo.org/sites/default/files/2024-11/GWR2024_English_ExecutiveSummary_WEB.pdf

Iqbal A, Haider R, Yasar A, Nizami A-S (2025) A governance model for sustainable municipal solid waste management: aligning the sector with Pakistan's economic goals. Waste Manag Bull 3(1):107–127. https://doi.org/10.1016/j.wmb.2024.12.010

Johannesen E (2025) Advancing gender equality in international ocean science: participatory approaches for institutional actions. Mar Policy 171:106421. https://doi.org/10.1016/j.marpol.2024.106421

Jolaosho TL, Rasaq MF, Omotoye EV, Araomo OV, Adekoya OS, Abolaji OY, Hungbo JJ (2025) Microplastics in freshwater and marine ecosystems: occurrence, characterization, sources, distribution dynamics, fate, transport processes, potential mitigation strategies, and policy interventions. Ecotoxicol Environ Saf 294:118036. https://doi.org/10.1016/j.ecoenv.2025.118036

Kumar M, Choubey VK, Raut RD, Jagtap S (2023) Enablers to achieve zero hunger through IoT and blockchain technology and transform the green food supply chain systems. J Clean Product 405:136894. https://doi.org/10.1016/j.jclepro.2023.136894

Larsen IB (2025) Project-based learning in business and management education: a scoping review and research agenda. Int J Manag Educ 23(2):101159. https://doi.org/10.1016/j.ijme.2025.101159

Magris RA, Marta-Almeida M, Monteiro JAF, Ban NC (2019) A modelling approach to assess the impact of land mining on marine biodiversity: assessment in coastal catchments experiencing catastrophic events (SW Brazil). Sci Total Environ 659:828–840. https://doi.org/10.1016/j.scitotenv.2018.12.238

Maher R, Buhmann K (2019) Meaningful stakeholder engagement: Bottom-up initiatives within global governance frameworks. Geoforum 107:231–234. https://doi.org/10.1016/j.geoforum.2019.06.013

Martin W (2019) Economic growth, convergence, and agricultural economics. Agricult Econ 50(S1):7–27. https://doi.org/10.1111/agec.12528

Moharrak M, AlReshaid F, Park KM, Alsaber AR (2025) International hidden entrepreneurs: concealed partnerships in new venture formation in an emerging markets context. J Innov Knowl 10(2):100669. https://doi.org/10.1016/j.jik.2025.100669

Montiel I, Gallo PJ, Antolin-Lopez R (2020) What on earth should managers learn about corporate sustainability? A threshold concept approach. J Bus Ethics 162(4):857–880. https://doi.org/10.1007/s10551-019-04361-y

Moretti V, Corraini NR, Melo EL, Scherer MEG, Colmenero JC (2024) Progress towards the sustainable development goal 14 (life below water) in the context of Brazil: a multicriteria approach. Sustain Fut 8:100410. https://doi.org/10.1016/j.sftr.2024.100410

Narong DK (2025) Business and management education for sustainability: a state-of-the-art review of literature post-UN decade of education for sustainable development (2015–2023). Int J Manag Educ 23(2):101115. https://doi.org/10.1016/j.ijme.2024.101115

Naseri H, Waygood EOD, Patterson Z (2025) Household transportation lifecycle greenhouse gas emission prediction. Transp Res D Transp Environ 141:104660. https://doi.org/10.1016/j.trd.2025.104660

Nielsen BB, Wechtler H, Zheng L (2023) Disasters and international business: insights and recommendations from a systematic review. J World Bus 58(4):101458. https://doi.org/10.1016/j.jwb.2023.101458

OECD (2024) Medición y control de la sostenibilidad del turismo a escala regional en España. https://www.oecd.org/es/publications/medicion-y-control-de-la-sostenibilidad-del-turismo-a-escala-regional-en-espana_9f52ddd8-es.html

Ordonez-Ponce E, Khare A (2021) GRI 300 as a measurement tool for the United Nations sustainable development goals: assessing the impact of car makers on sustainability. J Environ Plann Manag 64(1):47–75. https://doi.org/10.1080/09640568.2020.1746906

Pilgrim K, Bohnet-Joschko S (2022) Corporate social responsibility on Twitter: a review of topics and digital communication strategies' success factors. Sustainability 14(24):16769. https://doi.org/10.3390/su142416769

Ramirez J, Angelino Velázquez D, Vélez-Zapata C (2022) The potential role of peace, justice, and strong institutions in Colombia's areas of limited statehood for energy diversification towards governance in energy democracy. Energy Policy 168:113135. https://doi.org/10.1016/j.enpol.2022.113135

Rittel HWJ, Webber MM (1973) Dilemmas in a general theory of planning. Policy Sci 4(2):155–169. https://doi.org/10.1007/BF01405730

Rukanova B, van Engelenburg S, Ubacht J, Tan Y-H, Geurts M, Sies M, Molenhuis M, Slegt M, van Dijk D (2023) Public value creation through voluntary business to government information sharing enabled by digital infrastructure innovations: a framework for analysis. Govern Inform Quart 40(2):101786. https://doi.org/10.1016/j.giq.2022.101786

Rygh A, Chiarapini E, Segovia MV (2021) How can international business research contribute towards the sustainable development goals? Crit Perspect Int Bus 18(4):457–487. https://doi.org/10.1108/cpoib-08-2020-0123

Scheyvens R, Banks G, Hughes E (2016) The private sector and the SDGs: the need to move beyond 'business as usual.' Sustain Develop 24(6):371–382. https://doi.org/10.1002/sd.1623

Shaikh MA, Hadjikakou M, Geyik O, Bryan BA (2024) Assessing global agri-food system exceedance of national cropland limits for linking responsible consumption and production under SDG 12. Ecol Econ 215:107993. https://doi.org/10.1016/j.ecolecon.2023.107993

Shepherd B, Stone SF (2017) Trade and women. https://www.econstor.eu/handle/10419/163147

Shonk K (2025) Business conflict management

Sinha A, Shahbaz M (2018) Estimation of environmental Kuznets curve for CO_2 emission: role of renewable energy generation in India. Renew Energy 119:703–711. https://doi.org/10.1016/j.renene.2017.12.058

Sinkovics N, Vieira LM, van Tulder R (2022) Working toward the sustainable development goals in earnest: critical international business perspectives on designing and implementing better interventions. Crit Perspect Int Bus 18(4):445–456. https://doi.org/10.1108/cpoib-05-2022-0059

Sreenivasan A, Suresh M (2023) Exploring the contribution of sustainable entrepreneurship toward sustainable development goals: a bibliometric analysis. Green Technol Sustain 1(3):100038. https://doi.org/10.1016/j.grets.2023.100038

Sullivan K, Thomas S, Rosano M (2018) Using industrial ecology and strategic management concepts to pursue the sustainable development goals. J Clean Product 174:237–246. https://doi.org/10.1016/j.jclepro.2017.10.201

Tamasiga P, Onyeaka H, Bakwena M, Kayembe B, Dzingai V, Kgengwenyane N, Babugura AA, Ouassou EH (2025) Sustainable futures: aligning climate actions and socio-economic justice through the just energy transition. Energy Strat Rev 59:101726. https://doi.org/10.1016/j.esr.2025.101726

Ulpiani G, Vetters N, Shtjefni D, Kozarev V, Dunlop T, Pereira AG (2025) Do ambitious cities value collaboration in climate action? Insights from the first group of cities pursuing climate neutrality. Cities 162:105945. https://doi.org/10.1016/j.cities.2025.105945

United Nations Development Programme (2020) Sustainable development goals. www.undp.org/content/undp/en/home/sustainable-development-goals.html

United Nations Global Compact (2025) Ensuring a living wage is an essential aspect of decent work. https://unglobalcompact.org/what-is-gc/our-work/livingwages

Urata S, Narjoko DA (2017) Trade and inequality. WIN–WIN, 175

Vandenberg P (2017) Can trade help achieve the employment targets of the sustainable development goals? https://www.econstor.eu/handle/10419/163149

Van Dorp M, Martin M, Bojicic-Dzelilovic V (2025) Assessing peace and social impacts through local human security business partnerships. Bus Horiz. https://doi.org/10.1016/j.bushor.2025.03.004

Wang S, Bai X, van der Heijden J, Tong X (2024) The evolving roles of actors in sustainability experiments: evidence from community waste management in a Chinese city. Technol Forecast Soc Change 205:123469. https://doi.org/10.1016/j.techfore.2024.123469

World Business Council for Sustainable Development (2024) Integrating nature: assessing interconnected risks in the food retail ecosystem. https://www.wbcsd.org/resources/integrating-nature-assessing-interconnected-risks-in-the-food-retail-ecosystem/

World Trade Organization (2018) Mainstreaming trade to attain the sustainable development goals

Zhou Z, Yang Z, Li H, Liu J (2025) Regional climate change action and corporate ESG performance: evidence from China. Int Rev Econ Finance 100:104081. https://doi.org/10.1016/j.iref.2025.104081

Intangible Innovation: Non-Fungible Token (NFT) Price Evolution and Real-World Applications

Aldo Alvarez-Risco and Shyla Del-Aguila-Arcentales

Abstract Non-fungible tokens (NFTs) represent unique and indivisible assets on a blockchain, fundamentally differentiating themselves from fungible cryptocurrencies. Each NFT, backed by a smart contract, ensures authenticity and traceability of ownership, allowing creators to earn royalties and fostering vibrant secondary markets through interoperability. Their popularity skyrocketed during the COVID-19 pandemic, becoming a global trend with notable interest in various regions. Although the NFT market exhibits unique volatility and inefficient pricing, it is evolving beyond speculative digital art. NFTs are increasingly finding tangible, real-world applications, such as selling event tickets or academic diplomas, transforming into exploitable opportunities. Analysis of key collections such as Milady Maker, Pudgy Penguins, Bored Ape Yacht Club, Mad Lads, NodeMonkes, and Froganas reveals a dynamic, multi-chain ecosystem. These cases illustrate both the high-risk, high-reward nature of the market, with rapid appreciations and significant controversies, and a crucial shift toward functional utility, exemplified by executable NFTs (xNFTs). This study contributes to the understanding of NFT pricing and its practical applications, laying the groundwork for future market research.

Keywords NFTs · Blockchain · Prices · Cryptocurrency · Collectibles · Smart contracts

---

A. Alvarez-Risco (✉)
Universidad Tecnológica del Perú, Lima, Perú
e-mail: c27408@utp.edu.pe

S. Del-Aguila-Arcentales
Escuela de Posgrado, Universidad San Ignacio de Loyola, Lima, Perú
e-mail: sdelaguila@usil.edu.pe

A. Alvarez-Risco et al. (eds.), *Green International Business*, Environmental Footprints and Eco-design of Products and Processes,
https://doi.org/10.1007/978-3-032-12497-5_4

1 Introduction

Non-fungible tokens (NFTs) are a form of cryptocurrency that represents a single or indivisible asset on a blockchain (Turki et al. 2023). Unlike conventional cryptocurrencies, which are fungible and can be exchanged with each other on equal terms, NFTs are unique and cannot be exchanged in the same way. This means that each NFT has unique characteristics that make it irreplaceable and valuable in its own right (Ante 2022). The uniqueness of NFTs derives from their registration on the blockchain, which is a decentralized and public database that records all transactions of a particular cryptocurrency (Hawashin et al. 2024). Each NFT has its own unique digital identity and is backed by a smart contract, which is a computer code that sets the rules and conditions of the transaction. This technology guarantees the authenticity and exclusive ownership of the asset represented by the NFT; also, NFTs have become especially popular in recent years due to their ability to represent a wide variety of digital and physical assets, including digital art, music, videos, games, virtual real estate and other collectibles (Tavares et al. 2023). This has opened up new opportunities for content creators, artists and collectors to monetize their artworks and digital assets in innovative ways.

One of the most prominent features of NFTs is their ability to track the ownership history and authenticity of an asset over time. This means that every time a transaction is made with an NFT, it is recorded on the blockchain, allowing users to verify the ownership and origin of the asset transparently and securely (Divya et al. 2023); in addition, NFTs also allow for the inclusion of royalties in smart contracts, meaning that content creators can receive a commission every time their work is resold on the secondary market. This provides an ongoing form of income for artists and creators, even after they have initially sold their work. Another essential feature of NFTs is their interoperability, which means that they can be transferred between different platforms and ecosystems without losing their value or authenticity. This has led to the emergence of secondary marketplaces where users can easily buy, sell and exchange NFTs (van Haaften-Schick and Whitaker 2022).

During the pandemic caused by COVID-19, NFTs exploded in popularity as they became trending on the internet, especially in 2022 (Aharon and Demir 2022; Maouchi et al. 2022). The increased interest in NFTs can be seen in Google Trends (Fig. 1).

It is relevant to note which countries have the most interest in NFTs, such as China, Hong Kong, Singapore, Nigeria, South Korea, Philippines, Taiwan, Georgia and the United Arab Emirates (Google Trends 2025). NFTs initiated on the Ethereum network and then developed on other networks, such as Flow Blockchain, Solana, Cardano, or Binance Smart Chain, are purchased with the respective blockchain currency. Dowling (2021) showed that the pricing of NFTs appears to be different from the pricing of cryptocurrencies in terms of volatility transmission; also, another feature that is present in the NFT market is that pricing occurs inefficiently (von Wachter et al. 2023).

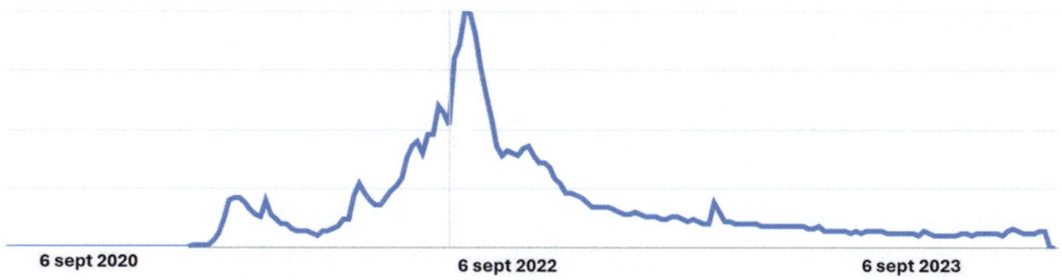

Fig. 1 According to Google's web search, the behavior of interest in NFTs during 2020–2023. *Source* Google Trends (2025)

Following the rise in popularity of NFTs, the end of August 2021 saw the highest secondary market sales volumes in the history of NFTs, reaching over $300 million daily (NonFungible 2021). While these levels have not been reached, an upward trend in NFT sales has been reported in early 2024. On the other hand, NFTs can also be helpful in the real world; as stated by Regner et al. (2019), it is possible to tokenize digital goods and represent physical assets that can be used in day-to-day life, for example, in the sale of event tickets or academic diplomas of a famous businessman or CEO. The NFTs exhibit varying price patterns, making it relevant to evaluate the historical information of this variation due to the innovation in NFT use. The current research contributes to enriching the literature on NFT pricing. It provides an analysis of NFT pricing as of the date of this study, addressing how NFTs are converted into tangible opportunities that can be exploited in the real world. Therefore, this research aims to establish the foundations for future research on this market.

2 Methodology

The foundation of our study lies in a comprehensive dataset that covers all transactions executed on secondary markets. Specifically, we focused on the ten NFT marketplaces that consistently record the highest weekly sales volume, allowing us to capture the most relevant and representative activity in the ecosystem. This selection ensures that the data reflects the trends and dynamics of the most active and liquid segments of the NFT market. For each NFT, the market capitalization, historical minimum and maximum price, and trading volume. The integration of these metrics provides a perspective on the performance of digital assets. The reliability of our data is basic, so the primary information is sourced from coingecko.com, a platform recognized for its comprehensiveness and accuracy in cryptocurrency and NFT data. Additionally, for detailed sales information on key NFT collectibles, we use specialized sources such as cryptoslam.io and nansen.ai, which offer in-depth analysis and transaction data at the collection level. This combination of sources ensures a solid and verifiable database for the conclusions of this study.

3 Results

3.1 NFT Milady Maker

This NFT is a distinctive 10,000-piece digital art collection, known for its neochibi style and cultural street influences (OpenSea 2024). The Remilia Collective created the collection, blending cutting-edge art with digital innovation. Since their launch in March 2021, they have gained prominence in the crypto community, embraced by influencers and integrated into various online platforms. In 2022, Milady Maker faced a major controversy that affected public perception and market value. Accusations arose against Charlotte Fang, known on Twitter as @CharlotteFang77 and CEO of the company (Jumpstart 2022). She was accused of being an anonymous user called "Miya", linked to starting an online misogynistic campaign against women. Other accusations pointed to Charlotte's involvement in the online exploitation of minors, whom she allegedly forced to self-harm. These serious allegations led to a sharp decline in the market value of Milady's NFTs, with the minimum price dropping dramatically from 1.5 ETH to 0.34 ETH. The situation began to stabilize when Charlotte Fang distanced herself from Milady Maker and publicly apologized for her past actions. This move was part of the company's efforts to address the controversy and rebuild its reputation in the NFT community. Figure 2 shows the transactions made in the secondary markets with the highest sales volume per week. The market cap is 29,884 ETH (July 31st, 2025). The maximum price was 7.035 ETH (December 2nd, 2024).

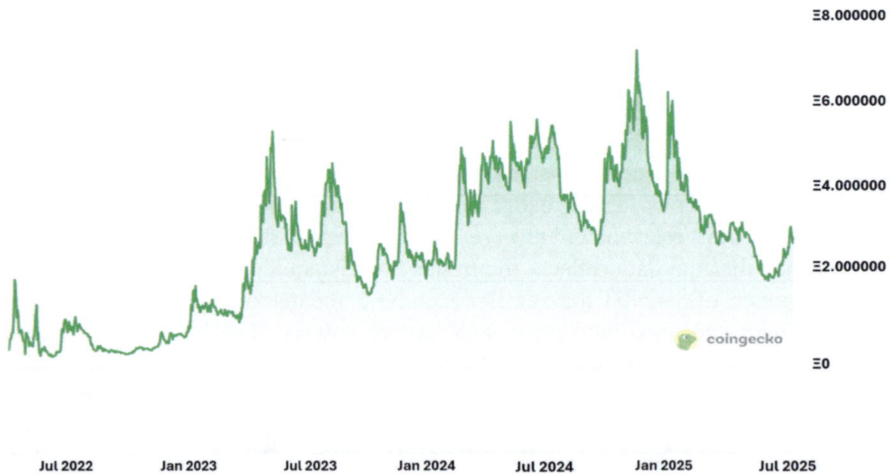

Fig. 2 Milady maker price trend. *Source* https://www.coingecko.com/en/nft/milady-maker

3.2 Pudgy Penguins

It is an influential NFT collection launched in July 2021, with 8888 unique tokens on the Ethereum blockchain. Beyond digital art, it encompasses a community-driven ecosystem with diverse content and merchandise. With over 266,000 ETH in trading volume, its market presence is significant (Phantom 2024). For example, chubby penguins with distinctive features such as golden, frosty or navy blue fur are highly sought after, with some fetching prices of up to 30 ETH. One of the most notable transactions in this collection includes the sale of Pudgy Penguin #6873, which fetched an impressive 400 ETH in a transaction in July 2021 (Crypto.com 2025). This information reflects the dynamic and often substantial market value associated with these digital collectibles. The collection includes the Lil Pudgys series, which adds to its appeal. Owners benefit from specific intellectual property rights, which allow for personal and some commercial uses (Crypto.com 2024). As of December 2022, the intellectual property rights to the Fat Penguins that come with ownership of an NFT grant the owner a limited, worldwide, non-exclusive license to use the associated artwork for personal, non-commercial use, and for commercial use under certain conditions (pudgypinguins.com 2022). Commercial use is permitted up to a gross revenue of $500,000 per year, unless the creator grants a broader license or an exemption. The market cap is 140,430 ETH (July 31st, 2025). The maximum price was 36.33 ETH (December 17th, 2024) (Fig. 3).

Fig. 3 Pudgy penguins price trend. *Source* https://www.coingecko.com/en/nft/pudgy-penguins

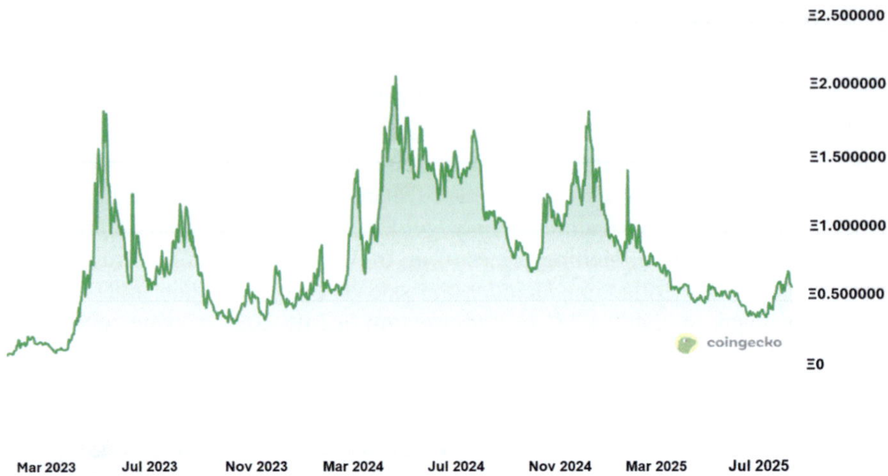

Fig. 4 Redacted remilio babies price trend. *Source* https://www.coingecko.com/en/nft/redacted-remilio-babies

3.3 Redacted Remilio Babies

It is a collection, hosted in Remilios, of 10,000 neochibi pfpNFTs expanding the Milady Maker paradigm with the introduction of young J.I.T. energy and schizophrenic reactionary aesthetic (Rarible 2024). Remilios forcefully swept the timeline with an outburst of adolescent energy, harnessing these inspirations to break away from the uniformity of NFT and crypto-twitter communities. Milady stands out as a symbol of directed chaos, embodying Remilia's revolutionary posting as an art form. The lucid metairony that has always been the essence of Remilia's practice sets her apart from the monotony of the online world. The market cap is 5982 ETH (July 31st, 2024). The maximum price was 2.122 ETH (April 22nd, 2024) (Fig. 4).

3.4 Ordinal Maxi Biz (OMB)

OMB's triadic collections and disruptive narrative make it unique. Led by zk-Shark, the project is committed to innovation and community empowerment, positioning it as a force with long-term potential. The market cap is 138.12 BTC (July 31st, 2025). The maximum price was 1479 000 BTC (March 7th, 2024) (Fig. 5).

Fig. 5 Ordinal Maxi Biz (OMB) price trend. *Source* https://www.coingecko.com/en/nft/ordinal-maxi-biz-omb

3.5 The Bored Ape Yacht Club

It is a collection of non-fungible tokens built on the Ethereum blockchain. It consists of 10,000 personalized ape drawings, each one being a unique variation of the same cartoon (Bloomberg 2022). This was described by Jimmy Fallon and Paris Hilton in their talk on NBC a year ago. Those who purchase the exclusive NFT from Yuga Labs will not only own the digital image but also gain access to a private online club. Yuga Labs recently reached a valuation of 4 billion dollars in March 2022 after closing a financing round of 450 million dollars (Cointelegraph 2022). NFTs were certainly not affordable for everyone. In January 2022, Fallon had bought his for $220,000, and Eminem had gotten another for almost $462,000 (Cointelegraph 2021). Having a colorful jumpsuit was a matter of status. As the months have passed, the chronicle of Bored Ape Yacht Club has been marked, however, by something more than cameos in prime time, since at the end of 2022, a lawsuit filed in Los Angeles by investors who felt deceived was revealed. In their sights, they had not only the company but a good number of celebrities (Observer 2022). The plaintiffs argue that the interest and value of NFTs have been artificially increased due to promotion with celebrities. This strategy led to digital works being purchased at inflated prices. The lawsuit promoters claim that the celebrities who participated in the campaign did not reveal that they had received compensation in return (Observer 2022). The market cap is 12,875 ETH (July 31st, 2025). The maximum price was 153.7 ETH (May 1st, 2022) (Fig. 6).

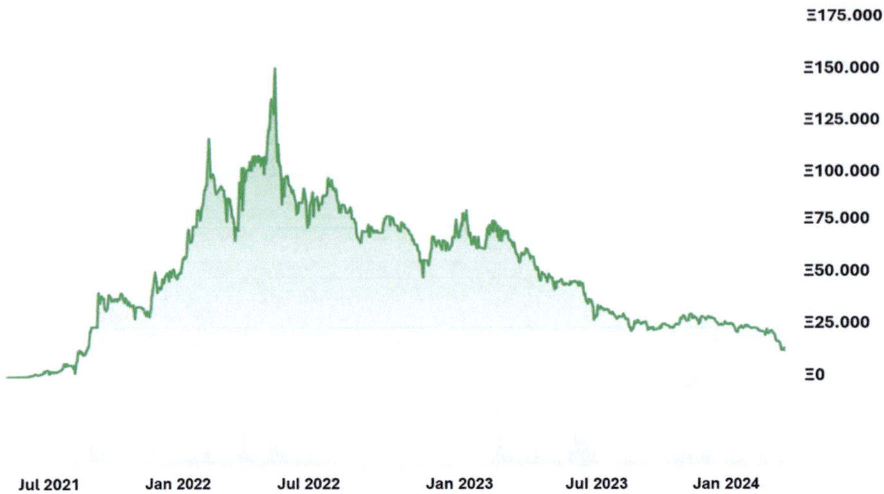

Fig. 6 Bored Ape Yacht Club price trend. *Source* https://www.coingecko.com/en/nft/bored-ape-yacht-club

3.6 *Mad Lads*

This collection is from Backpack, an open-source multi-chain crypto wallet available as a Chrome extension. It is worth mentioning that Backpack was developed by Coral, the team dedicated to the development of the so-called xNFTs. The X in xNFTs stands for eXecutable (executable code), which represents the ownership rights over its execution. An xNFT can be a game, an NFT gallery, or a cryptocurrency price tracker—anything that can be written with executable code (Spozz 2023). Mad Lads NFT in its first days on the market. The founders announced that they had achieved 7879 unique holders after the announcement. The next day, April 22, sales had surpassed $5 million. The emergence of Mad Lads generated a significant impact, causing a collection on the Solana network to top the list with more than 8000 transactions on its first day. A week after its launch, the floor price of the Mad Lads collection was 89 SOL. A particularity of this collection is that its founders sought to achieve a high number of unique owners (Spozz 2023). The market cap is 371,643 SOL (July 31st, 2025). The maximum price was 229.425 SOL (Dec 9th, 2023) (Fig. 7).

3.7 *NodeMonkes*

It burst onto the scene in early 2023, featuring a collection of 10,000 distinctive images permanently inscribed on the Bitcoin blockchain. These images, ranging

Fig. 7 Mad lads price trend. *Source* https://www.coingecko.com/en/nft/mad-lads

from whimsical monkeys to enigmatic ducks, are created by compressing strange-looking blocks, producing unique characters. The project, led by an anonymous team, auctioned 8000 of these records in December 2023, with a dynamic pricing model that captivated bidders from around the world. NodeMonkes is a collection of 10,000 Bitcoin Ordinals inscriptions. Although the collection was inscribed in February 2023, its developers waited until December before releasing it to the public (CCN 2024). In a collectors' culture that values older block stamps, the Ordinals' claim to be the first 10,000-count profile picture collection inscribed on Bitcoin counts for a lot. NodeMonkes have amassed a market capitalization of $567 million (Coinlive 2024). Between February 29 and March 4, the market cap more than doubled to 8850 BTC. Not only are NodeMonkes now the most valuable Ordinals NFTs, they are also worth more than Pudgy Penguins, Mutant Ape Yacht Club and other popular Ethereum-based collections. The collection achieved another significant milestone on Monday, March 4, when one of the inscriptions, NodeMonke #2769, sold for 17 BTC. This was the first time that one of the inscriptions exchanged hands for over a million dollars (Binance 2024). The market cap is 616 BTC (July 31st, 2025). The maximum price was 0.895 BTC (Mar 4th, 2024) (Fig. 8).

3.8 Froganas

The Solana ecosystem has recently experienced an upswing with the emergence of very profitable ventures like the Meem Token, as well as a significant surge in the NFT market over the past few days. One of the most fascinating NFT projects in

Fig. 8 NodeMonkes price trend. *Source* https://www.coingecko.com/en/nft/nodemonkes

this ecosystem is Froganas. Froganas was released on December 22nd at a price of 0.069 SOL with a total supply of 5555 units, and was sold out within a swift 2-min window. Tensor, a Solana trading market, reported that Froganas has reached a floor price of 14.47 SOL today, boasting a staggering surge of over 14,000% from its minting price. However, this project faced a premature demise before achieving such immense success. It was the perseverance of its creators that resurrected Froganas from its demise. The market cap is 4395 SOL (July 31st, 2025). The maximum price was 8.93 SOL (Jan 24th, 2024) (Fig. 9).

3.9 Mutant Ape Yacht Club (MAYC)

It is a collection of 20,000 NFTs that features artwork of mutant monkeys. It was created on August 28, 2021, as the second expansion of the Bored Ape ecosystem. The purpose of MAYC is to attract more collectors to the Bored Ape ecosystem. Since its launch, the collection has become extremely popular and is now the leader in the PFP NFTs market (Binance 2024). Prior to the launch of the Mutant Ape Yacht Club (MAYC), there were only two popular profile image (PFP) NFT projects in the market—Bored Ape Yacht Club (BAYC), which launched in April 2021, and Bored Ape Kennel Club, which launched in June 2021. As the demand for PFP NFTs increased during the summer of 2021, developers were seeking new and innovative ways to launch their avatar projects. Yuga gained attention when it introduced MAYC with unique mechanics (Coleccion NFT 2023). Half of the 20,000 NFTs within the MAYC collection were created this way, and the other half were sold through auctions. The auction started at 3 ETH (around $10,000 at the time) and was set to

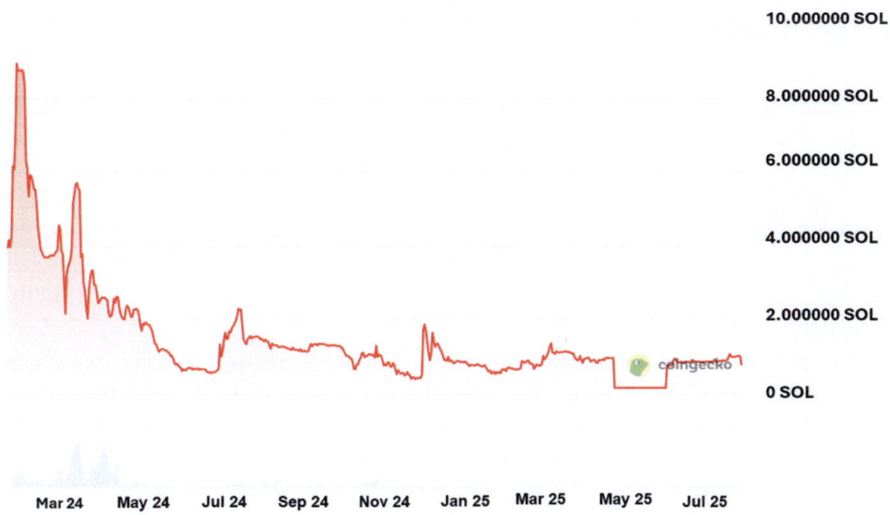

Fig. 9 Froganas price trend. *Source* https://www.coingecko.com/en/nft/froganas

drop to 0.01 ETH over nine hours. However, the auction lasted only about an hour, reaching a final price of about 2.79 ETH, and all the Mutant Ape NFTs were minted (NFT Now 2022). The market cap is 35,780 ETH (July 31st, 2025). The maximum price was 40.0 ETH (May 1st, 2022) (Fig. 10).

Fig. 10 Mutant Ape Yacht club price trend. *Source* https://www.coingecko.com/en/nft/mutant-ape-yacht-club

4 Discussion

This study focuses on the evolution and intricacies of the non-fungible token (NFT) market. The evolution of the NFT market, characterized by its uniqueness and indivisibility on the blockchain, has been assessed. Unlike fungible cryptocurrencies, each NFT possesses properties that make it irreplaceable, guaranteeing its authenticity and exclusive ownership through smart contracts and immutable records on the blockchain. This allows for transparent tracking of ownership history and facilitates the inclusion of royalties for creators, ensuring a continuous stream of income in the secondary market.

The interoperability of NFTs, which allows them to be transferred across various platforms without loss of value, has played a key role in their growth, particularly during the COVID-19 pandemic, becoming a global trend in 2022. This increased interest is reflected in Google Trends data, with notable peaks in countries such as China, Hong Kong, Singapore, Nigeria, and South Korea, suggesting diverse and geographically widespread adoption, with unprecedented sales volumes reached by the end of August 2021. While these peaks have not been sustained, an upward trend in sales was observed in early 2024, indicating a potential market maturation beyond the initial speculative euphoria.

The price dynamics of NFTs compared to conventional cryptocurrencies are striking, particularly with regard to volatility transmission. This means that the value of NFTs is derived from fluctuations in the underlying cryptocurrencies and is influenced by intrinsic factors such as artistic merit, utility, or cultural relevance. Furthermore, the NFT market is characterized by pricing inefficiency, meaning prices do not always incorporate all available information. This inefficiency, while potentially creating opportunities for sophisticated trading strategies, also carries significant risks for less-informed investors, a common characteristic in emerging markets that lack mature price discovery mechanisms. NFT price patterns vary, emphasizing the need to review history for understanding their development and uses. Besides being digital collectibles, NFTs have practical applications like tokenizing digital and physical assets, such as tickets or diplomas. Analyzing key collections, including Milady Maker and Bored Ape Yacht Club (BAYC), shows market diversity and risks. Controversies, like allegations against Milady Maker's CEO causing a price drop, and lawsuits accusing BAYC of inflated values through undisclosed promotions, demonstrate the market's vulnerabilities, ethical issues, and the importance of transparency.

Collections like Pudgy Penguins highlight how NFTs can foster community-driven ecosystems and confer intellectual property rights on their holders, enabling limited commercial uses. The rise of Mad Lads on Solana, with its concept of xNFTs (executable NFTs), marks a significant step toward functional utility, where an NFT can function as a game, arcade game, or price tracker. NodeMonkes, permanently registered on the Bitcoin blockchain and quickly reaching a substantial market capitalization, exemplifies the market's expansion across different blockchains and the creation of high-value assets. The rapid exit and return of Froganas on Solana, despite its untimely demise, demonstrates the resilience of the community and the high-risk, high-reward nature of the market. The proliferation of prominent collections on platforms like Ethereum, Solana, and Bitcoin suggests a multi-chain future and fierce competition for NFT dominance. This platform diversity impacts transaction costs and speed, and signals a shift from simple profile pictures to executable, IP-protected NFTs, emphasizing functional utility beyond digital art.

5 Conclusion

The NFT market has rapidly evolved from a niche digital collectible scene to a key part of the digital economy. Enabled by blockchain technology and smart contracts that ensure authenticity and ownership, its popularity emerged, especially following the COVID-19 pandemic, reflecting increased interest in digital asset monetization and new ownership models. Despite this growth, the market still shows pricing inefficiencies and distinctive volatility, indicating its early development stage. A notable shift toward practical utility is emerging, with applications like ticketing, diploma sales, and NFTs with specific features such as xNFTs. The broad range of collectibles across blockchains like Ethereum, Solana, and Bitcoin suggests a multi-chain future and fierce competition. While offering high-risk, high-reward opportunities, the market faces controversies and regulatory hurdles, emphasizing the need for greater transparency and accountability. For sustained long-term growth and broader adoption, clearer regulations and enhanced market efficiency will be essential, facilitating better integration of NFTs into traditional financial systems and promoting their use for different activities.

References

Aharon DY, Demir E (2022) NFTs and asset class spillovers: lessons from the period around the COVID-19 pandemic. Financ Res Lett 47:102515. https://doi.org/10.1016/j.frl.2021.102515

Ante L (2022) The Non-Fungible Token (NFT) market and its relationship with bitcoin and ethereum. FinTech 1(3):216–224. https://doi.org/10.3390/fintech1030017

Binance (2024) NodeMonke #2769 Sold for 17 BTC, Valued at Approximately $1.08 Million. https://www.binance.com/es-MX/square/post/2024-03-04-nodemonke-2769-sold-for-17-btc-valued-at-approximately-1-08-million-4938072542273. Accessed on 03/03/2024

Bloomberg (2022) Greg Solano and Wylie Aronow, Bored Ape Universe Creators. https://www.bloomberg.com/news/articles/2022-12-14/greg-solano-wylie-aronow-bored-ape-universe-creators-bloomberg-50-2022. Accessed on 03/03/2024

CCN (2024) NodeMonkes Bitcoin ordinals collection climbs past pudgy penguins in NFT charts. https://www.ccn.com/news/nodemonkes-bitcoin-ordinals-collection-surpasses-pudgy-penguins/#:~:text=NodeMonkes%20is%20a%20collection%20of,million%20dollars%20worth%20of%20BTC. Accessed on 03/03/2024

Coinlive (2024) NodeMonkes has a market capitalization of $558 million, surpassing Pudgy Penguins and MAYC. https://www.coinlive.com/news-flash/456236. Accessed on 03/03/2024

Cointelegraph (2021) Bored Ape Yacht Club is a huge mainstream hit, but is Wall Street ready for NFTs? https://cointelegraph.com/news/bored-ape-yacht-club-is-a-huge-mainstream-hit-but-is-wall-street-ready-for-nfts. Accessed on 02/02/2024

Cointelegraph (2022) Yuga Labs, creador de NFT, recauda USD 450 millones y eleva la valoración de la empresa a USD 4.000 millones. https://es.cointelegraph.com/news/nft-creator-yuga-labs-raises-450m-bringing-company-valuation-to-4b. Accessed on 02/02/2024

Coleccion NFT (2023) Mutant Ape Yacht Club: La colección complementaria de BAYC. https://www.coleccionnft.es/mutant-ape-yacht-club/. Accessed on 03/03/2024

Crypto.com. (2024) Lil Pudgys. https://crypto.com/price/es/nft-collections/lilpudgys-1?blockchain=0. Accessed on 03/03/2024

Crypto.com (2025) What Is Pudgy Penguins (PENGU)? https://crypto.com/en/university/what-is-pudgy-penguins-pengu. Accessed on 20/07/2025

Divya P, Rajeshwaran S, Parthiban R, Ananth Kumar T, Jayalakshmi S (2023) Hashgraph: a decentralized security approach based on blockchain with nft transactions. In: Bhattacharya A, Dutta S, Dutta P, Piuri V (eds) Innovations in data analytics. Springer Nature Singapore, pp 33–50. https://doi.org/10.1007/978-981-99-0550-8_3

Dowling M (2021) Fertile LAND: pricing non-fungible tokens. Financ Res Lett 102096. https://doi.org/10.1016/j.frl.2021.102096

Google Trends (2025). NFT trends between 2021 and 2023. Retrieved 03/03/2024 from http://trends.google.es/trends/explore?date=2020-01-01%202023-12-31&q=NFT&hl=es

Hawashin D, Nemer M, Salah K, Jayaraman R, Svetinovic D, Damiani E (2024) Blockchain and NFT-based traceability and certification for UAV parts in manufacturing. J Ind Inf Integr 39:100597. https://doi.org/10.1016/j.jii.2024.100597

Jumpstart (2022) Milady maker—a look into the seedy underbelly of NFT art. https://www.jumpstartmag.com/milady-maker-a-look-into-the-seedy-underbelly-of-nft-art/. Acceseed on 02/02/2022

Maouchi Y, Charfeddine L, El Montasser G (2022) Understanding digital bubbles amidst the COVID-19 pandemic: evidence from DeFi and NFTs. Financ Res Lett 47:102584. https://doi.org/10.1016/j.frl.2021.102584

NonFungible (2021) The world's largest NFT data resource. https://nonfungible.com. Acceseed on 02/02/2022

NFT Now (2022) A pillar of the BAYC universe: meet the mutant ape yacht club. https://nftnow.com/guides/a-pillar-of-the-bayc-universe-meet-the-mutant-ape-yacht-club/. Acceseed on 03/03/2024

Observer (2022) Justin Bieber and Madonna are among the celebrities sued over promoting bored ape NFT*s*. https://observer.com/2022/12/justin-bieber-and-madonna-are-among-the-celebrities-sued-over-promoting-bored-ape-nfts/. Acceseed on 03/03/2024

OpenSea (2024) Milady maker. https://opensea.io/collection/milady. Acceseed on 03/03/2024

Phantom (2024) Pudgy Penguins is pushing NFTs into the mainstream. https://phantom.app/explore/nft-spotlight/pudgy-penguins. Acceseed on 03/03/2024

pudgypinguins.com. (2022) IPRights. https://pudgypenguins.com/ip-rights. Acceseed on 02/02/2022

Rarible. (2024) Redacted remilio babies. Retrieved 02/02/2024 fromhttps://rarible.com/remilio-babies/items. Acceseed on 02/02/2024

Regner F, Schweizer A, Urbach N (2019) NFTs in practice-non-fungible tokens as core component of a blockchain-based event ticketing application. https://www.fim-rc.de/Paperbibliothek/Veroeffentlicht/1045/wi-1045.pdf. Acceseed on 12/22/2021

Spozz (2023) Mad Labs: what is behind the NFTs that have surprised the market? https://blog.spozz.club/es/todo/tendencias/mad-lads-que-hay-detras-de-los-nfts-que-han-sorprendido-al-mercado/. Acceseed on 03/03/2024

Tavares R, Sousa JP, Maganinho B, Gomes JP (2023) Gamers' reaction to the use of NFT in AAA video games. Procedia Comput Sci 219:606–613. https://doi.org/10.1016/j.procs.2023.01.329

Turki M, Cheikhrouhou S, Dammak B, Baklouti M, Mars R, Dhahbi A (2023) NFT-IoT pharma chain: IoT drug traceability system based on blockchain and non fungible tokens (NFTs). J King Saud Univ Comput Inform Sci 35(2):527–543. https://doi.org/10.1016/j.jksuci.2022.12.016

van Haaften-Schick L, Whitaker A (2022) From the Artist's Contract to the blockchain ledger: new forms of artists' funding using equity and resale royalties. J Cult Econ 46(2):287–315. https://doi.org/10.1007/s10824-022-09445-8

von Wachter V, Jensen JR, Regner F, Ross O (2023) NFT wash trading. In: Matsuo SI, Gudgeon L, Klages-Mundt A, Perez Hernandez D, Werner S, Haines T, Essex A, Bracciali A, Sala M (eds) Financial cryptography and data security. FC 2022 international workshops. Springer International Publishing, pp 299–311. https://doi.org/10.1007/978-3-031-32415-4

Developing Global Environmental Management and Sustainability: The Case of a German Multinational "Hidden Champion" in Brazil

Madeleine Bausch

Abstract Amid growing environmental and social awareness, multinational companies (MNCs) are under increasing pressure to integrate sustainability into their global operations and value chains. While sustainability and corporate social responsibility (CSR) are often a burdensome part of compliance for many MNCs, there are fewer companies that genuinely regard sustainability as a driver of innovation and competitive advantage for global market development. This chapter presents the case study of Aqualith, a German multinational hidden champion that has successfully integrated environmental management into its global corporate quality strategy. As a pioneer in sustainability issues, the company views environmental management not as a regulatory obligation, but as a driver of long-term value creation. Based on interviews and field research in Germany and Brazil, the case analysis focuses on how Aqualith defines and manages its environmental management, transfers it to its Brazilian subsidiary, and adapts it to local conditions. The case highlights three crucial facets of global environmental management at Aqualith: (1) the strategic integration of environmental management and sustainability standards into the global quality management approach, (2) the balance between global consistency and local adaptation, thereby overcoming institutional, cultural and organizational challenges, and (3) its contribution to the development of sustainability in Brazil as an emerging market. Finally, this chapter derives eight practical implications of global environmental management that can help MNCs go beyond regulatory compliance and leverage sustainability as a strategic advantage.

Keywords Environmental management · Sustainability · Global strategy · MNC · Germany · Brazil · Emerging markets

M. Bausch (✉)
Department of Business Administration, Faculty of Economics and Business, University of Chile, Santiago, Chile
e-mail: mbausch@fen.uchile.cl

A. Alvarez-Risco et al. (eds.), *Green International Business*, Environmental Footprints and Eco-design of Products and Processes,
https://doi.org/10.1007/978-3-032-12497-5_5

1 Introduction

Sustainability has turned into an indispensable topic that companies, especially multinational companies (MNCs), must deal with. Pressures from governments, society, and competitors are driving companies to redesign their strategies, structures and processes to meet the responsible, sustainable, and inclusive demands of their stakeholders. For example, in 2023, the German government has adopted the Supply Chain Act based on the EU Corporate Due Diligence Directive, which is intended to improve the protection of human rights and the environment in the global supply chains of German companies that source globally. From 2024 on, all companies with more than 1,000 employees are required to ensure compliance with this law in relation to their direct suppliers and to report on it (The Federal Government of Germany 2024).

In the realm of global business, corporate social responsibility (CSR) and environmental sustainability initiatives are typically developed within advanced economies and subsequently transferred by MNCs to their foreign subsidiaries (Aguilera et al. 2021; Kolk 2016). While these companies often excel in implementing these strategies in their home countries, where corporate practices align closely with domestic regulations and stakeholder expectations, they frequently encounter difficulties when attempting to replicate these efforts abroad (Asmussen and Fosfuri 2019; Bausch et al. 2022). This challenge is particularly pronounced in subsidiaries operating in emerging economies (Ghauri 2018; Jamali et al. 2020; Yin and Jamali 2016), where institutional frameworks and cultural contexts can differ significantly. Host-country conditions such as weak CSR regulations, institutional voids, and distinct cultural values or interpretive systems regarding environmental and social protection (Bausch 2022; Preuss et al. 2024; Wehling et al. 2009) can then be major barriers to the effective transfer and implementation of CSR practices. In such environments, aligning actual behavior with corporate rhetoric becomes increasingly difficult, making it a significant challenge for firms to genuinely "walk the talk" of their global CSR commitments. For instance, the Nestlé Group, headquartered in Switzerland, has faced criticism for failing to align its global CSR commitments, particularly those targeting plastic waste reduction, with practices in Global South countries, where recycling infrastructure remains underdeveloped (Swissinfo 2020). Additionally, environmental disasters such as oil spills caused by Western multinationals like Royal Dutch Shell, Texaco, and Chevron in regions like the Niger Delta (Jolaoso 2025) and the Ecuadorian Amazon (BBC 2018) underscore the severe consequences of inadequate and inconsistently applied CSR strategies by Western firms operating in developing contexts.

Moreover, while large MNCs can more easily comply with these demands by establishing dedicated CSR departments, small and medium-sized enterprises (SMEs) and small MNCs often face significant challenges in doing so (Park and Ghauri 2015). Limited financial and human resources hinder their ability to implement and maintain the required standards, and their international suppliers are frequently too small to comply effectively. As a result, sustainability can become a burdensome obligation rather than a strategic avenue for innovation and long-term value creation. Consequently, many companies aim to merely satisfy the minimum legal requirements, with few incentives in place to pursue meaningful sustainability initiatives. In this context, CSR and sustainability efforts risk becoming a "ceremonial facade" (Meyer and Rowan 1977, p. 358), driven more by regulatory oversight than by intrinsic, bottom-up motivation.

Although government initiatives and conceptual frameworks like the Triple Bottom Line, ESG (Environmental, Social, and Governance), and the United Nations Sustainable Development Goals (SDGs) have elevated the global conversation on the importance of integrating sustainability into business practices, many MNCs now place disproportionate emphasis on demonstrating strategic alignment with these frameworks. This often results in a performative focus on compliance—such as showcasing ESG scores or SDG alignment—without fully embedding sustainability into the core of their business models and day-to-day operations. The sustainability and "circular economy" initiatives promoted by fast-fashion giants like H&M (Ellen MacArthur Foundation 2021; H&M Group 2024) and Shein (2025) highlight best the tension between the strategic image projected and the actual sustainability performance of their operational business models.

While most MNCs today have formal CSR strategies in place, only a minority truly view sustainability as a catalyst for innovation, a source of long-term competitive advantage and long-term returns, and a foundation for global market development. Embedding CSR across the entire value chain—from idea generation to production and distribution, rather than limiting it to marketing—can drive genuine sustainable transformation and reshape the definition of long-term economic success and profitability. By integrating sustainability into innovation, firms can gain a competitive advantage, expanding the concept of "success" beyond financial performance to include meaningful societal impact. Conversely, when CSR efforts are disconnected from actual practices, they risk not only environmental harm with long-term ecosystem consequences but may also damage a firm's reputation and social legitimacy in the long run. This chapter therefore contends that a broader shift is urgently needed: more companies must adopt sustainability not as a compliance exercise or marketing narrative, but as an integral part of their corporate DNA if they aim to realize a meaningful transformation toward sustainable development.

This chapter presents an illustrative case study of Aqualith,[1] a German multinational that has successfully embedded sustainability into its global corporate strategy. Going beyond mere compliance with domestic regulatory requirements, Aqualith has positioned sustainability and CSR as a key driver of innovation and long-term value creation for a wide range of stakeholders. Notably, the company has also demonstrated an ability to effectively transfer its sustainability practices across borders. The case thus illustrates that investments in sustainability and environmental protection can yield tangible returns, effectively challenging the widespread assumption that sustainability and profitability are mutually exclusive. This chapter focuses on the Aqualith's environmental management approach and sustainability initiatives

[1] The real company name was anonymized for this study.

and their successful implementation in its Brazilian subsidiary. Employing a single-case study methodology, the analysis seeks to answer the central question: *How does Aqualith define, manage and effectively transfer its environmental management system to its foreign subsidiary in Brazil?*

The chapter is structured as follows: Section two presents the description of the company and highlights its unique characteristics that make it worth being studied as an exemplary single case for sustainability and environmental management initiatives. Section three outlines the methodology used to explore the research question. Section four then presents the findings, followed by a list of practical implications for other globally operating companies and business leaders. Key recommendations include, for example, integrating environmental management into firm strategy and operations, balancing standardization with local adaptation, and using bicultural translation dyads to strengthen intercultural communication to achieve cohesion between company sites.

2 The Case

Aqualith is a German multinational "hidden champion" from Stuttgart (located in the federal state of Baden-Württemberg), which is active in the industrial technology sector. A "hidden champion" company is relatively unknown to end-customers due to limited media exposure, and a global leader in its niche market though strong expertise and high innovation capacity (Simon 2009). As a business-to-business company, its core business embraces control systems for technological firms of multiple branches, such as automotive, construction, or medical industry. Besides, with its business unit Aqualith Training, it is a pioneer in industry education and offers technical education services for schools, universities, and companies.

The company was founded in the 1920s. Its legal form is a limited partnership with non-family top managers serving the Board of Management; several family members serve the Supervisory Board. The thrive for innovation and sustainability has always been at the heart of the company. With a strong inspiration from nature, this organization develops solutions that are inspired by biology, physics and chemistry. More than 7% of its revenue is invested in research and development, such as a collaborative network that brings together universities and industry partners to explore product innovation from interdisciplinary perspectives. Additionally, through its training initiatives, the organization fosters technical learning and vocational training at universities.

In line with its connection to nature and its value for longevity, the company is a pioneer in corporate sustainability issues, not only from a strategic, but also from an operative point of view. On a strategic level, environmental management and sustainability are firmly anchored in its corporate strategy and reflected in its corporate structure. Apart from being certified according to ISO14001 (international standard for environmental management), the company is committed to the environment, for example, through their development of an in-house sustainability strategy, its care for ecological footprint in product development, and its strategic orientation toward a steady reduction of CO2 emissions. On an operative level, concrete measures include the tracking of the groups' carbon emissions, its energy, water, and wastewater use, power generation through photovoltaic systems, and the use of material-efficient product design. Global production sites are further equipped with energy monitoring systems, executed and controlled by local "energy teams."

Importantly, the company not only cares about their own carbon footprint, but also about the carbon footprint of their suppliers and clients: the company is therefore concerned with inventing and producing products that help clients to reduce CO_2 energy efficiency, such as through modules for energy efficient production. Moreover, Aqualith offers consultation services to achieve CO2 transparency, long-term use instructions of products and related materials, and gives advice on the energy-optimized use of Aqualith products for its clients. For its achievements in sustainability, Aqualith has been awarded several Sustainability Prizes.

Globally, the company is organized in 11 country clusters. Aqualith is active in over 60 countries and has more than 200 subsidiaries. By 2025, it employs around 20,000 employees and has 30,000 customers in 180 countries. In Brazil, Aqualith initiated its activities through a Greenfield production subsidiary in the 1960s, when the Brazilian economy opened for Foreign Direct Investment. Sales and production units are located in the city of São Paulo. The plant produces foremost for the Latin American market. The Brazilian subsidiary employs around 200 people and is exclusively staffed with local managers. Aqualith Brazil was one of the first companies in Brazil to achieve ISO9001 certification (international standard for quality management systems) in the 1990s.

The research for this study was conducted within the "Corporate Quality and Environment" department at Aqualith's German headquarters in Stuttgart, as well as at its Brazilian subsidiary in São Paulo. At the headquarters' level, approximately 40 employees are engaged in quality-related responsibilities, with a dedicated team of five focusing exclusively on sustainability. This team is responsible for overseeing the global implementation of Aqualith's sustainability strategy and the dissemination of related practices across its international production sites. At the Brazilian subsidiary, this department is integrated into the production function and is divided into two segments: "Quality Management Systems and Environment," which focuses on environmental and management aspects, and "Quality Management and Logistics," which addresses operational quality concerns. The quality team at the Brazilian subsidiary is considerably smaller, comprising only five members in total. Within this team, one individual is solely responsible for environmental issues.

3 Method

This chapter adopts a single-case study approach (Ghauri 2004; Yin 2017), focusing on a company that represents a particularly compelling example of success in sustainability and environmental management. Single-case studies are especially well-suited for generating in-depth insights into complex phenomena, allowing for a detailed examination of the distinctive characteristics, strategies, and practices (Yin 2017). In-depth insights from single-case studies are thus well-suited to highlight exemplary business practices that can inspire other firms, while also revealing pitfalls and missteps that serve as lessons. Furthermore, case studies enable rich contextualization, capturing the phenomenon within the specific temporal and spatial settings in which it unfolds, thereby offering a more holistic understanding (Marschan-Piekkari and Welch 2011).

3.1 Data Collection

The data was collected as part of a field study conducted between June 2018 and April 2019 at the German headquarters and the Brazilian subsidiary. One of the major strengths of case-based field research is the combination of multiple data sources through triangulation, to provide deep insights into the phenomena under investigation (Yin 2017).

This study draws on four data sources: semi-structured interviews, focus groups, documents, and field notes, which were later developed into analytical memos. 25 semi-structured interviews were conducted with managers and employees from various departments, including Quality and Environment, Research and Development, Human Resources, and Top Management. Interviews were held in German, Portuguese, and one in Spanish, lasting between 30 min and 1.5 h. The interview guide included questions about the interviewees' roles and backgrounds, the structure and tools of quality and environmental management, and the strategic and operational transfer of related practices between Germany and Brazil. A key focus was placed on how practices were adapted within the Brazilian context, alongside the challenges, success factors, and intercultural dynamics influencing communication and collaboration.

Two focus groups were further conducted with employees from the Brazilian subsidiary's quality department to explore the local dynamics of implementing quality and environmental initiatives. These sessions, supported by interactive techniques such as Metaplan cards, offered insights into how these practices were conceptualized and how elements of corporate culture manifested within the subsidiary.

Moreover, 18 documents complemented the primary data. These included internal company records, protocols from initial meetings, relevant sections of the corporate website, and practitioner-authored articles, offering additional context and triangulation.

Lastly, during her field research, the researcher took notes and photos on the spatial and temporal aspects of the interviews. This included observations of buildings and research sites, notes on factory tours, participation in meetings, lunches and coffee breaks, as well as spare time with participants. The field notes, summarized into eight larger memos, helped to contextualize the research, and to strengthen arguments from interviews, focus groups, and documents.

3.2 Data Analysis

The interviews were recorded with a recording device, after asking for consent among the interviewees, and transcribed in the original languages German, Portuguese, and Spanish. All transcripts, field notes and collected documents were then digitalized and inserted into the computer-aided coding and data analysis program MaxQDA 2024 (Kuckartz and Rädiker 2019). Due to the uniqueness of the case, this study adopts an inductive data analysis procedure, yet, with a-priori concepts in mind (Miles et al. 2014). These concepts included existing frameworks of sustainability, such as the triple bottom line (Elkington 1997) and ESG (Environmental, Social, Governance), as well as on the concepts of global integration and local responsiveness (Bartlett and Ghoshal 1998), transfer of sustainability practices (Gutierrez-Huerter et al. 2020; Jamali et al. 2020), as well as ideas on organizational and national development (Neumayer and Perkins 2005). The richness and triangulation of data sources strengthened the validation of individual interview statements and enabled cross-checking with documents and field notes, enhancing the robustness of the data analysis framework.

4 Findings

The case of Aqualith reveals three overall facets of global environmental management: (1) the global strategic and operative integration of environmental management through cross-cultural transfer, (2) the balance between global consistency and local adaptation of environmental management to overcome institutional, cultural, and organizational differences, and (3) its contribution to the development of sustainability in Brazil as an emerging market through education and certification initiatives.

4.1 Global Strategic and Operational Integration of Environmental Management Through Cross-Cultural Transfer

From a strategic point of view, Aqualith integrates its environmental management and sustainability approach, which is based on the ESG framework, into its overall corporate strategy:

> Sustainability is embedded in our corporate strategy. As a business with a long-term perspective, we embrace our responsibility for environmental, social, and governance (ESG) issues to benefit both current and future generations. For us, sustainability entails giving equal and comprehensive consideration to all three dimensions. (Document 11)

The organization's sustainability strategy is structured around four main pillars: Planet Stewardship, Integrity and Responsibility, People and Wellbeing, and Knowledge for Innovation. These pillars are aligned with the Sustainable Development Goals: Planet Stewardship covers responsible energy and water use, a commitment to reach net-zero emissions by 2030, and efforts to reduce material consumption. People and Wellbeing emphasizes employee growth, workplace safety, and diversity. Knowledge for Innovation advances sustainable progress through technical education and learning. Integrity and Responsibility ensures compliance, human rights due diligence, and ethical supply chain practices. This chapter focuses on the first pillar, Planet Stewardship, highlighting how environmental management strategy and practices have been transferred to one of the organization's international subsidiaries, Brazil.

The International Quality and Environmental Management System

To anchor its sustainability strategy in its global corporate processes, Aqualith set up its "International Quality and Environmental Management System" (IQEMS) in 2005, which combines the characteristics of a Quality Management System with its Environmental Management System. The company therefore integrates the ISO standards 9001 (international standard for quality management systems), 14001 (international standard for environmental management), and OHSAS, now termed ISO45001 (international standard for health and safety management in the workplace) in one system (Norbert, Manager Germany; Tales, Employee Brazil; Jefferson, Employee Brazil). In its essence, the IQEMS works as a preventive system. A Quality Manager at the headquarters compares the system to an "insurance" that keeps quality at a certain level: "It's always a protection against product liability. This is often used as an argument because no one wants to go to jail or be guilty of something" (Felix, Manager Germany), where "something" relates to violations of environmental, governance, or social integrity. The Brazilian quality manager also assures that "there's also the environmental aspect that we maintain, to ensure that we are not doing anything harmful in terms of the environment." (Felipe, Manager Brazil).

In contrast to many other firms that often define a Quality Management System based on the ISO9001 standard, the IQEMS at Aqualith considers quality holistically, as it involves environmental safety and sustainability, employee well-being and governance integrity: "this quality management system is not as narrowly defined as it is at many companies, that it is purely focused on quality systems. […] It has a broader focus" (Michael, Employee Germany). The company therefore defines quality in three key dimensions: product, process and corporate quality (document 6; Focus Group 2).

The cross-cultural transfer of the IQEMS to Brazil

The transfer of the IQEMS to Brazil started in 2006, when the headquarters "decided that we should extend this Environmental Management System to our foreign sites if we were to produce more internationally in future" (Norbert, Manager Germany). To bring the IQEMS to life, Aqualith has broken down the system into a corporate process map that defines group standards and corporate processes:

> Over the past 13 years, we've integrated all production, logistics, and development units into a Global Quality and Environmental Management System. A group certificate with corporate procedures applies to all 14 sites worldwide, including Brazil. (Felix, Manager Germany)

> Aqualith Germany sets the standards and tools for evaluating quality, which all sites must follow. In terms of environmental quality, this includes understanding environmental aspects, their impacts, and how to minimize them. (Jefferson, Employee Brazil)

The cross-cultural transfer of environmental management primarily involved implementing the IQEMS using the Corporate Process Map in its global subsidiaries, including Brazil (Ana, Employee Brazil; Felipe, Manger Brazil; Nico, Employee Germany). In accordance with ISO standards, Aqualith's overarching corporate process map comprehensively covers functions such as development, production, procurement, sales and environmental management. A highly detailed process map reflects Aqualith's strong focus on process standardization and quality, and clearly outlines the roles, responsibilities and mutual obligations of "internal customers" to ensure consistent quality (Felix, Manager Germany; Luis, Supervisor Brazil; Hugo, Employee Brazil). This quality consistency is then operationalized through three core methods (Norbert, Manager Germany): a method for identifying and evaluating environmental aspects to derive targeted measures and objectives; a method for continuous environmental monitoring and control; and a method for ensuring compliance with legal environmental requirements, incorporating them into internal processes, and regularly monitoring and evaluating adherence.

At the operational level, environmental management at the headquarters and the Brazilian subsidiary is manifested through several key practices: proactive waste management, the implementation of energy and water monitoring systems through energy teams, and frequent discussions and training on the ISO14001 certification as a benchmark for environmental performance.

Waste separation emerged as a particularly important topic at the Brazilian subsidiary, extending beyond the production floor into office spaces. During her fieldwork, one of the first observations the researcher made upon entering the production facilities was the colored waste bins positioned at the entrance of the office buildings. A closer inspection revealed that employees were adhering to the waste separation guidelines—a notable contrast to the practices commonly observed in São Paulo's city center, where waste is often indiscriminately mixed despite the presence of sorting bins. Moreover, as well inside the buildings, employee adhered to waste separation, as they differentiated bins for non-recyclables, paper, and plastics.

Second, the Brazilian subsidiary set up water and energy monitoring systems in the form of an internal control method. Even though German environmental protection standards are said to be stricter than in Brazil, it is especially the São Paulo region which has long-standing experience in dealing with water shortages: "São Paulo itself has little water. Companies must make sure that water consumption becomes a critical issue at a relatively early stage." (Norbert, Manager Germany). The Brazilian Manager Felipe reassures: "São Paulo is very strict in terms of the environment." A local employee, invented an in-house "internal control instrument", which ultimately was re-transferred to Germany where it was considered a best practice for the whole group:

> We have over ten environmental indicators, like water and energy use. I set up a system where each person enters data directly into SharePoint for automatic calculation. Germany used to do this manually, but Norbert adopted the idea and created a central portal, which I now also use. (Jefferson, Employee Brazil)

Aqualith Germany compares these indicators on a monthly basis across all sites to measure each plant's water and energy consumption. To encourage each plant to reduce water and energy consumption, the company has set up an internal competition system and ranks the most environmentally friendly subsidiaries. An "environmental committee" (Jefferson, employee in Brazil), which includes all the people involved in environmental issues, is responsible for tracking consumption levels.

For Aqualith Brazil, the control tool "brought a lot of results for our indicators" (Larissa, Manager Brazil), as it had practical consequences for waste and water reduction:

> We had a water-chilling unit using 200,000 liters annually. I pushed for a change but was initially blocked due to cost. After showing the expenses—water, energy, and maintenance—I convinced the company to rent a new unit, which the board approved last month. (Jefferson, Employee Brazil)

Third, the importance of environmental management becomes evident through a frequent mentioning of the ISO14001 norm in headquarters and subsidiary (Ana, Employee Brazil; Felipe, Manager Brazil; Tamara, Employee Germany):

> These are our two main tasks: to establish, implement and further develop a certifiable ISO14001 system in the global production network, and to address environmental protection concerns within the company. (Norbert, Manager Germany)

Especially in Brazil, interview partners highlight moreover the importance of the OHSAS certification:

> We make health and safety a very high priority, including OHSAS. I think Aqualith is one of the very few companies here in Brazil that has OHSAS. We've always focused a lot on that part of health and safety. (Felipe, Manager Brazil)

Taken together, these practices demonstrate that Aqualith successfully aligns its environmental management strategy at both headquarters and the subsidiary with their operational practices. More than just strategic integration, cross-cultural transfer has allowed the company to effectively embed sustainability into its daily operations and routines.

4.2 *Localized Adaptation While Maintaining Global Consistency*

While Aqualith sets its global environmental standards through a top-down corporate strategy and its corporate process map, it also engages local subsidiaries in dialogue to allow for contextual adaptation. This balanced approach reflects its "glocalization" strategy—combining global consistency with local responsiveness—and reinforces its identity as both a "global player and independent business" (Document 11) that respects local culture while maintaining a unified global vision:

> Think globally but act locally. Acting locally, of course, has provided the flexibility and great autonomy of action that has allowed the organization to grow, allowing the company to expand to the sixty or so countries in which Aqualith operates today. (Augustin, Managing Director Brazil)

Aqualith pursues a clear global consistency, as "regarding quality and environment, we have one common management system worldwide" (Norbert, Manager Germany). The global IQEMS is implemented across all corporate units to unify subsidiaries under one system. To ensure compliance, the headquarters conducts annual corporate audits to check

> whether the system has been effectively implemented. In other words, whether the processes are broken down completely, correctly and unchanged to the local plant. (Felix, Manager Germany)

While Aqualith ensures alignment by transferring corporate standards, it also allows flexible adaptations at the Brazilian subsidiary. Adaptation therefore serves as a mean to overcome institutional, cultural, and organizational challenges that the company faces due to its embeddedness in different national contexts.

As German manager Felix explains, "the transfer of standards, support, networking… all happens with the involvement of the local organization," emphasizing that it's "not just one way; 'headquarters knows everything better'." Local implementation is therefore shaped through exchange, taking institutional, cultural and organizational factors into account.

First, several voices highlight the "integration" (Joachim, Manager Germany) of the Brazilian plant in shaping the IQEMS, particularly in adapting procedures and environmental standards to local laws and regulations:

> We review and adapt requirements based on local conditions, starting with country-specific laws. Each site's quality team operates within the local legal entity and reports to the plant manager or managing director. (Felix, Manager Germany)

> We set processes to ensure sites follow local environmental regulations. For example, while we define waste management broadly, each country must handle disposal according to its own laws—German waste law can't be applied internationally. The Brazilians have to collect and dispose of their waste according to Brazilian waste law. [...] We don't specify anything internationally. (Norbert, Manager Germany)

Instead of designing a global IEQMS to fit in all countries where Aqualith operates, the company leaves such adaptations to the subsidiary, not least because the Brazilian legislation is one of the most complex in the world (Canuto and Ribeiro dos Santos 2018). Evidence for localization efforts is also backed by Brazilian interviewees: "When a process is not applicable, we don't have to implement it here" (Jefferson, Employee Brazil); "We have the freedom to discuss with Germany, so that we can say it doesn't work that way here, and we can do it differently" (Larissa, Manager Brazil).

Second, while headquarter and subsidiary employees recognize the difficulties that arise due to cultural differences, such as the degree of openness when talking about errors, the degree of initiative-taking on behalf of the workforce, or the frequency of communication (Tamara, Employee Germany; Hugo, Employee Brazil; Anna, Employee Brazil), the company manages to overcome these cultural differences through mutual adaptation and communication. One the one hand, some processes are adapted to local culture, such as the Brazilian "jeitinho," a cultural concept that describes the creative circumvention of rules and laws when people feel they do not fit the context (Duarte 2006):

> We can modify the process depending on the procedure. If it's mandatory, we must find a jeitinho to comply with it. If it's not mandatory, but there's a difference in Brazilian legislation, we need to adapt it. (Anna, Employee Brazil).

> We don't follow it exactly as it is. It may be because of the lack of analytical profile, or because of the hurry itself; I want to do everything faster. Brazilians want to make things faster and easier. Dar to jeitinho and make things faster. I think that in this point we end up sinning a little bit. (Larissa, Manager Brazil)

The *jeitinho* thus encompasses adaptation, but also the importation and creative invention of new practices, whenever needed. Subsidiary employees import environmental practices from other companies where they have worked. For example, having previously worked at another European company, Larissa noticed the issue of individual wastebaskets at desks, which created waste without separation. She then centralized the bins and introduced waste separation, aligning with the company's environmental philosophy (Larissa, Manager Brazil).

Third, Aqualith uses its local adaptation strategy to overcome organizational differences between headquarters and subsidiary. One issue is the structural disbalance between the numbers of employees working at the headquarters, and the number of employees at the subsidiary. While the headquarters employs approximately 40 employees in CSR-related tasks, the team at the Brazilian subsidiary comprises only five members. Since a 1:1 transfer would create a heavy workload problem at the subsidiary, the local quality team could "redesign" the corporate process map, simplifying it to make it fit the lesser number of Brazilian employees: "I have to put some [activities] together, reduce the activities. [...]. And we can do that. But I can't leave doing the activities that are crucial." (Larissa, Manager Brazil). As the Brazilian plant unites productions and sales within one company, the Brazilian map also includes both production and sales processes, in difference to the German one (Felix, Manager Germany; Felipe, Manager Brazil). Moreover, to make documents fit the employee structure in Brazil, the subsidiary develops own documents (Felipe, Manager Brazil), and designs context-adapted corporate procedures (Anna, Employee Brazil).

Key Drivers to Balance Standardization and Localization

Three key drivers lead to an effective balance of localized adaptation and global consistency, overcoming institutional, cultural, and organizational challenges.

First, Aqualith appoints "local and global process responsible" employees for environmental management at the subsidiary, making it not only a practice "everyone" should meet, but a practice that is actively managed by a group of responsible people. Respectively, Aqualith creates "local projects" where responsible persons for quality are appointed in cooperation with top management (Felix, Manager Germany). At the headquarters, environmental management is headed by Norbert, Environmental Manager, who, in his own words, has a "passion for environmental protection." Therefore, he pushed environmental issues forward by initiating a project for eco-certification and integrating environmental standards into the corporate quality norm. At the subsidiary, Jefferson is responsible for environmental issues:

> At first, I joined to do accident prevention work. But because the area needed a coordinator, I ended up being responsible for OHSAS. Then I ended up working on the environment and ISO14001. (Jefferson, Employee Brazil)

Clear responsibilities foster close cooperation and communication, not only on a professional, but also on a personal level (Jefferson, Employee Brazil).

Second, bicultural translation dyads served as a key mechanism to effectively transfer the IEQMS, balancing standardization and localization. Bicultural translation dyads are defined a team of individuals from the involved cultures—Germany and Brazil—who collaborate to communicate knowledge and practices across borders (Bausch 2022). At Aqualith, these bicultural translation dyads are characterized by very close, personal and "extremely proximate relationships" (Jefferson, Employee Brazil). For example, the German managers and employees Felix, Norbert, and Tamara are denominated as proper "Brazilians" by Brazilian interview partners (Anna, Employee Brazil; Jefferson, Employee Brazil), even though none of them grew up in Brazil or speaks fluently Portuguese:

> Norbert has this quality of hugging people [in the figurative sense]. Germans are usually a little less approachable, especially at work. Brazilians drink coffee with friends every day. The company is not only a place of work, but also a place of friends. I have the impression that Germans are a little less like that. But when Norbert came to Brazil two weeks after my daughter was born, he brought a gift and wanted to see a photo. He came with a teddy bear in his hand, and I thought that was a very Brazilian attitude. So I think also because of the years of experience they have at Aqualith, they are closer to us. (Jefferson, Employee Brazil)

This closeness facilitates the communication and collaboration between Jefferson at the subsidiary, and Norbert at the headquarters, to transfer sustainability-related processes and procedures between the two entities.

Third, the company is constantly working on and developing its organizational culture, strengthening cooperation between local and global managers. Regular intercultural communication between headquarters and subsidiary staff balances globalization and localization efforts. Employees emphasize the "climate of cooperation" and the sincere interest of the headquarters in the development of the subsidiary, which creates a sense of family. This is reflected in the ongoing mutual negotiations and exchanges. The Brazilian employees agree that "the headquarters supports us very well" (Larissa, Manager Brazil), and the interviewees emphasize the "proximity" (Focus Group 1; Focus Group 2; Sérgio, Employee Brazil), "respect" (Focus Group 1) and "trust" (Felipe, Manager Brazil; Jefferson, Employee Brazil) in the headquarters-subsidiary relationship:

> The respect is very great. It ends up being like a big family. They always try to work on this climate issue based on respect, based on everyone's understanding. Transforming the company into a big family. (Focus Group 2)

> In recent years, Germany has shown a strong desire to be closer, which is great. We receive many invitations, often too many to attend. The Germans' focus on sharing is highly productive and aligns well with the Brazilian culture. (Adrian, Manager Brazil)

This intercultural communication and exchange, paired with familial proximity meets the Brazilian people-oriented culture, supports an adaptation of practices while maintaining global standards, and thus entails an open attitude toward new practices and corporate innovation.

4.3 Development of Environmental Management and Sustainability in Brazil

The transfer of the IQEMS marks a significant step for Aqualith's Brazilian subsidiary, not only establishing a foundational environmental management system but also catalyzing a wider sustainability transformation within Brazil. This far-reaching impact is being realized through dedicated employee training programs, comprehensive vocational training, and proactive supplier development and certification efforts, extending sustainable practices throughout the value chain and to key stakeholders.

Development of Environmental Management at the Brazilian Subsidiary

The transfer of the IQEMS achieves the further development of the subsidiary in terms of environmental management, in particular through training measures. These training measures include Aqualith's in-house Quality Consciousness Trainings, which promote quality and environmental thinking among employees at global plants and aim to "promote the necessary awareness and provide basic and advanced know-how" on quality (Document 7). The training includes information on the company's quality standards, the company's quality philosophy and related behavioral expectations of the workforce, understanding the impact of non-quality, and awareness of the needs of internal customers.

The transfer of the IQEMS, paired with the training measures, leads to a gradual alignment between headquarters and subsidiary environmental standards and processes. While the Brazilian plant previously had its own tools and practices for assessing and evaluating environmental indicators, the transfer of the IQEMS required alignment of standards and processes:

> [There is] more standardization, to follow procedures and to be aligned with the headquarters. I have many designs here that I used to make modifications here in Brazil [...]. Not today. A product from Brazil has exactly the same specifications. (Felipe, Manager Brazil)

> Before, I had my own tool that we had developed to do the evaluation and registration. Then, Germany developed another tool and, because it is corporate, I must use it. We follow the same base of evaluation tools, forms, procedures that Germany follows. (Jefferson, Employee Brazil)

Training on the IQEMS content also led to a clearer understanding of process responsibilities and boundaries with their "internal customers" (Jefferson, Employee Brazil), i.e., the person or department within the organization who receives products, information, or support from another person or department:

> With the training, people started to think more about the process as a whole. Because until then they thought only about their work. But with the training they were able to think about the process, and what and who comes before them and what comes after, and what it is going to affect. (Anna, Employee Brazil)

As a result, the company is moving toward its holistic view of quality, in line with the company's overall environmental strategy:

> Before, quality was very plastered. [...] Quality was very much focused on the product, less on the process and the company. [...] Nowadays, the process is much better considered (Jefferson, Employee Brazil)

> Germany saw how seriously we worked in the last four years to fix many processes. Other areas didn't have described procedures or standards, and they didn't know how to design a procedure. (Larissa, Manager Brazil)

To sum up, Aqualith's innovative Quality Consciousness Trainings enable the company to achieve a gradual socialization and convergence of its quality concepts and understanding among the employees of the subsidiaries. This promotes environmental awareness among local employees, leading to the implementation of sustainability practices that have a direct impact on the subsidiary's environmental performance (Larissa, Manager Brazil; Martin, Manager Germany; Tamara, Employee Germany).

Development of Sustainability in Brazil through Training and Certification

Beyond its direct impact on developing environmental management at the subsidiary, Aqualith also contributes indirectly to developing sustainability and environmental awareness in Brazil as an emerging market. This is achieved through vocational training and local supplier development and certification.

Vocational training is one of the pillars of Aqualith's corporate strategy and is, wherever possible, adopted in local subsidiaries. Aqualith Brazil therefore cooperates with local educational institutions to provide technical courses, for example with SENAI (National Industrial Learning Service) to offer dual education and training:

> We have agreements with a large number of the country's universities. The institutions here are SENAI, various technical and vocational training institutions where we continue to collaborate, transfer knowledge and technology, provide training directly. [...] And we make a very important contribution, also with Germany, in dual education. We are always acting in these scenarios; with the Ministry of Education, the education policies themselves where we are part of and help in the construction of the curricular plans. (Agustin, Managing Director Brazil)

In doing so, the company emphasizes the inherent social and long-term perspective of educational measures that do not yield immediate, but long-term financial returns. Instead of focusing on profit, Aqualith reveals its future-oriented character:

> [Education] is an area of our natural work. Participating in education is not a business in itself, but it does sow opportunities for the future. [...] In the very long term. So, what you are harvesting was sown 12, 15, 20 years ago. (Agustin, Managing Director Brazil)

Moreover, local supplier development and certification serve as an important measure to enhance quality and environmental standards in Brazil. Due to Brazil's protective economy, Aqualith sources around 80% from local suppliers, and imports around 20% from Germany and international suppliers (Tales, Employee Brazil). To ensure and enhance supplier standards, Aqualith conducts regular audits. Supplier quality, including environmental quality, is therefore firmly anchored in the corporate quality concept (Tamara, Employee Germany; Martin, Manager Germany; Tales, Employee Brazil).

Whenever needed, Aqualith engages in trainings with local suppliers in which they adopt preventive and corrective supplier development methods measures and 8D reports, that is, a process for problem-solving that consists of eight steps to address non-conformities in manufacturing and supply (Martin, Manager Germany; Leonardo, Manager Brazil):

> When I see that the supplier has a chronic problem that needs development, I call the people from the Supplier Quality Development department. Now, if the supplier has a one-off problem, I'll work with the 8D. (Tales, Employee Brazil)

During this process, the company also assures that suppliers meet certification standards such as ISO9001 and ISO14001. As of the first companies in Brazil to achieve ISO9001 certification in 1993, Aqualith thus disseminates this certification standard beyond the firm to local suppliers. This leads to more quality and environmental awareness among direct Aqualith suppliers in Brazil:

> What we do have is strong supplier support. There are visits to the supplier by the purchasing staff. And then we have a kind of supplier certification, so that the suppliers deliver with the guaranteed quality. (Adrian, Manager Brazil)

In this way, compliance with sustainability standards has become more important not only for Brazilian suppliers of Aqualith, but also for Brazilian companies in general over time. This is, on the one hand, due to stricter Brazilian regulation:

> The dependence on ISO14000 or the environmental part in Brazil has become more rigid. I don't know about the rest of the country, but São Paulo is very strict in terms of the environment. (Felipe, Manager Brazil).

On the other hand, subsidiary employees also highlight the increased importance of environmental compliance among consumers as "Our clients constantly demand for environmental, health and safety and quality issues, too" (Jefferson, Employee Brazil).

> The customer is very interested in certificates. Whether it's a quality certificate or something that's environmentally certified. Because, as part of your product, when you buy a product, you expect it to be of high quality and not pollute the environment. (Tales, Employee Brazil)

In sum, through its education and certification measures within and beyond the firm, Aqualith marks a cornerstone for quality and environmental management importance in Brazil. Thus, the case not only illustrates how an MNC can successfully establish and diffuse sustainability practices and environmental management within its global corporate network, but also how it can extend and diffuse its standards and practices beyond the company to stakeholders in its local ecosystem.

5 Practical Implications

From the case study, this chapter derives eight practical implications that MNCs can follow to achieve more and honest environmental management and sustainability practices in their global operations.

1. **Strategic integration of sustainability and environmental management into the overall corporate strategy**

The case study on Aqualith highlights the importance of embedding sustainability and environmental management deeply within the corporate strategy. This can be done by either setting up an own-standing environmental strategy, or by integrating it into existing strategic frameworks, such as the overall corporate quality strategy. Doing so not only reinforces the company's core identity—what it stands for—but also provides a clear direction for its long-term objectives. A well-defined strategic framework enables companies to identify relevant areas of action and, just as crucially, to exclude those that fall outside their environmental scope. This focus sharpens goal-setting and enhances strategic coherence.

Moreover, established frameworks such as ESG (Environmental, Social, and Governance) criteria or the Triple Bottom Line offer orientation for organizations at the initial stages of strategic alignment. Additionally, the United Nations Sustainable Development Goals (SDGs) can support the formulation of more concrete and globally relevant objectives. Once key areas are selected, they should be translated into specific sub-goals that adhere to the SMART principle—Specific, Measurable, Achievable, Rewarding, and Time-bound. To ensure accountability and track progress, companies should also develop targeted Key Performance Indicators (KPIs) tied to their sustainability and environmental objectives.

2. **The importance of operational and structural integration to bring the strategy to life**

Yet, while embedding environmental management and sustainability into the corporate strategy is essential, it is not sufficient on its own. A purely strategic or even symbolic commitment may enhance corporate image but fails to generate tangible outcomes if not translated into everyday practice. To drive real impact, the strategy must be operationalized and integrated into daily business activities. Aqualith serves as a strong example of this approach, having effectively embedded environmental management within its International Quality and Environmental Management System. This system, aligned with ISO9001 and ISO14001 standards, outlines a comprehensive corporate process map and assigns specific roles and responsibilities across departments. Such structural integration ensures that sustainability is not an abstract ideal, but a concrete and actionable part of the organization's workflow.

Equally critical is the establishment of clearly defined responsibilities within the Environmental Management System. Designating individuals who are not only accountable but also intrinsically motivated to push environmental transformation is important. These committed individuals become drivers for change, making them indispensable to a functioning and effective Environmental Management System.

3. **Ensuring global consistency in environmental management**

To maintain a coherent and unified approach to environmental management across international operations, companies must establish clearly defined corporate strategies, standards, and processes. These should be effectively communicated to subsidiaries and local markets through deliberate and structured dissemination efforts. Tools such as translated documentation, in-person company visits, training programs, and educational initiatives play an important role in ensuring that corporate guidelines are both understood and implemented locally. In the absence of such resources, subsidiaries may develop their own materials and tools, which can lead to inconsistencies and dilute the company's global environmental identity. To avoid this, headquarters must take an active role in monitoring and supporting local implementation. This includes regular communication, auditing, and performance reviews to ensure alignment with corporate standards.

However, it's important to strike a balance: excessive oversight and control may undermine the autonomy and motivation of subsidiary employees. A more effective approach involves fostering a sense of shared ownership, providing guidance and support while allowing local units the flexibility to adapt within defined strategic parameters.

4. **Adapting environmental management to institutional, cultural, and organizational contexts**

As with any management system, Environmental Management Systems must be adaptable to local cultural, institutional, and organizational contexts. Countries vary not only in terms of environmental regulations, corruption levels, and reporting requirements, but also in how sustainability is perceived and prioritized by firms, suppliers, and customers. In many emerging markets, environmental standards are less stringent or even absent, and local businesses often face more pressing challenges, such as high employee turnover, a shortage of qualified personnel, unreliable supplier quality, political instability, or crime, that can push environmental concerns down the priority list.

In addition, significant differences in subsidiary structures must be taken into account. Corporate processes designed at headquarters—often in developed markets and within large, well-resourced organizations—may not be directly transferable to smaller subsidiaries with leaner staffing and simplified operations. To ensure effective implementation, headquarters must cultivate an understanding of the institutional, cultural, and organizational environments of their subsidiaries, particularly in emerging markets. This includes adapting strategies, structures, and processes to align with local realities, rather than imposing uniform systems that may not be practical or effective on the ground.

5. **Embedding global standards through communication and education.**

The successful transfer of global environmental management standards hinges on ongoing communication and continuous education efforts at the subsidiary level. Given that employees are shaped by diverse educational systems, the familiarity and maturity of quality and environmental concepts can vary significantly across regions. As such, targeted training becomes essential not only to explain the technical elements of an Environmental Management System, but also to convey the deeper principles and values behind it.

The case of Aqualith underscores the value of internal employee training and development programs as a means to communicate and reinforce the company's broader sustainability vision. Aqualith takes a long-term approach to employee development, recognizing that meaningful learning and behavioral change require time. The company invests in its workforce even when immediate financial returns are not evident, setting itself apart from many firms in Brazil and across Latin America, where high employee turnover often limits long-term development initiatives.

Beyond the direct benefits within the subsidiary, MNCs can leverage training and education to contribute to broader societal change. By collaborating with local schools, universities, and vocational institutions, companies can help build local capacity and promote the diffusion of environmental management principles in emerging markets.

6. **Fostering intercultural communication between headquarters and subsidiaries**

Aqualith has successfully cultivated an organizational culture grounded in mutual respect and trust, where employees—particularly those at the subsidiary level—feel genuinely heard and valued. This foundation is crucial for effective knowledge and practice transfer, as people are the conduits through which information flows. As such, intercultural communication and mutual understanding are core enablers of successful implementation.

This is especially important in person-oriented cultures, such as Brazil and many other emerging markets, where interpersonal relationships often take precedence over task execution. In these contexts, personal, face-to-face communication becomes a vital mechanism for building trust and ensuring the successful transfer of environmental management practices. By investing in meaningful intercultural dialogue and fostering human connection across organizational boundaries, companies can enhance both the consistency and effectiveness of global sustainability efforts (Barmeyer et al. 2021).

7. **Appointing bicultural translation dyads**

Intercultural communication flourishes when individuals with the appropriate cultural competencies are placed in key positions. The case of Aqualith illustrates that one of the central mechanisms for balancing standardization and localization within its Environmental Management System was the use of bicultural translation dyads—teams composed of individuals from both involved cultures, in this case Germany and Brazil, who jointly facilitate cross-border knowledge transfer (Bausch 2022; Bausch et al. 2022). These dyads typically comprise individuals with substantial cultural familiarity, gained through language proficiency, professional experience, personal ties, or time spent in the respective foreign context (Barmeyer and Davoine 2019). Such individuals may have previously lived abroad, worked for a foreign affiliate company, or maintained close familial or social connections across cultures. Effectively identifying these key actors often depends on robust knowledge management systems capable of capturing not only explicit, but also tacit cultural and experiential knowledge embedded within the organization.

8. **Active supplier development to strengthen and expand environmental standards**

The case study illustrates how Aqualith engages in proactive supplier development and certification to ensure compliance with environmental and quality standards. These practices not only improve the firm's own product quality and environmental footprint but also extend its influence beyond internal operations, helping to elevate standards among local suppliers. By establishing clear requirements such as supplier certification, foreign MNCs can actively shape and support the development of environmental management practices within their subsidiary networks and the broader local economy.

In this way, MNCs serve as standard-setters, potentially driving systemic improvements in environmental performance across regions. Certifications such as ISO9001 and ISO14001 play a crucial role in this process, offering internationally recognized benchmarks to define, monitor, and enhance standards. This is particularly valuable in emerging markets like Brazil, where local environmental regulations may still lag behind those of developed economies. By leveraging supplier development as a strategic tool, MNCs can contribute meaningfully to both corporate sustainability goals and wider economic, social, and environmental progress.

6 Conclusion

This chapter has presented the case of Aqualith, a German multinational "hidden champion" company, a forerunner in sustainability issues, and which has managed to transfer its Environmental Management System successfully to its overseas production subsidiary in Brazil. The case study has provided insights into how this MNC defines, manages, and transfers environmental management on a global level, transforming sustainability into an innovative tool.

Aqualith integrates its Environmental Management System, including sustainability goals in its global corporate Quality Management Strategy, defining an in-house International Quality and Environmental Management System (IQEMS). This system helps the company, beyond the integration of sustainability in its overall quality strategy, to define environmental protection as part of the quality it produces, and to anchor its environmental goals in its operative corporate processes. A clear definition of the system and its related processes facilitate the transfer of its system to its Brazilian subsidiary.

Moreover, the case revealed that Aqualith achieves a localized adaptation of its IQEMS while maintaining global consistency, thereby overcoming institutional, cultural, and organizational differences. This is achieved through intense local workforce training, the employment of bicultural translation dyads, and intercultural communication, creating an organizational culture of cooperation and trust. Such culture of cooperation, trust and respect seems especially important in the German-Brazilian context, with the Brazilian culture being marked by the importance of personal relationships.

Finally, this chapter has also shown how Aqualith's environmental management not only supports the development of environmental awareness at the Brazilian subsidiary but also promotes the spread of sustainability concepts beyond the subsidiary to Brazilian suppliers and stakeholders. In-house employee training, vocational training and local supplier development and certification serve as key drivers for a shift toward more environmental management and sustainability practices in their direct supplier network in Brazil.

In summary, the case reveals how Aqualith goes beyond mere compliance with international regulations to create a competitive advantage through sustainability and environmental management. The eight practical implications can inspire other local and multinational companies to further develop and improve their environmental strategies and sustainability-related actions.

References

Aguilera RV, Aragón-Correa JA, Marano V, Tashman PA (2021) The corporate governance of environmental sustainability: a review and proposal for more integrated research. J Manag 47(6):1468–1497

Asmussen CG, Fosfuri A (2019) Orchestrating corporate social responsibility in the multinational enterprise. Strateg Manag J 40(6):894–916

Barmeyer C, Davoine E (2019) Facilitating intercultural negotiated practices in joint ventures: the case of a French-German railway organization. Int Bus Rev 28(1):1–11

Barmeyer C, Bausch M, Mayrhofer U (2021) Constructive intercultural management: integrating cultural differences successfully. Edward Elgar Publishing, Cheltenham

Bartlett CA, Ghoshal S (1998) Managing across borders: the transnational solution. Harvard Business Press, Boston

Bausch M (2022) Intercultural transfer of management practices of German MNC to Brazil: the interplay of translation and recontextualization. Springer Nature, Wiesbaden

Bausch M, Barmeyer C, Mayrhofer U (2022) Facilitating factors in the cross-cultural transfer of management practices: the case of a German multinational in Brazil. Int Bus Rev 31(2):101921

BBC (2018) Chevron wins Ecuador rainforest 'oil dumping' case. BBC. Retrieved from https://www.bbc.com/news/world-latin-america-45455984. Accessed on 14 May 2025

Canuto O, Ribeiro dos Santos T (2018) Economic effects of the Brazilian constitution. Novos Estudos 37(3):417–426

Duarte F (2006) Exploring the interpersonal transaction of the Brazilian jeitinho in bureaucratic contexts. Organization 13(4):509–527

Elkington J (1997) The triple bottom line. In: Russo M (ed) Environmental management: readings and cases. Sage, Thousand Oaks, pp 49–66

Ellen MacArthur Foundation (2021) Designing products for circularity: H&M Group. Retrieved from https://www.ellenmacarthurfoundation.org/circular-examples/hm-group. Accessed on 14 May 2025

Ghauri P (2004) Designing and conducting case studies in international business research. In: Piekkari R, Welch C (eds) Handbook of qualitative research methods for international business. Edward Elgar Publishing, Cheltenham, pp 109–124

Ghauri P (2018) Multinational enterprises and sustainable development in emerging markets. In: Bergé JS, Harnay S, Mayrhofer U, Obadia L (eds) Global phenomena and social sciences: an interdisciplinary and comparative approach. Springer, Cham, pp 21–36

Gutierrez-Huerter OG, Moon J, Gold S, Chapple W (2020) Micro-processes of translation in the transfer of practices from MNE headquarters to foreign subsidiaries: the role of subsidiary translators. J Int Bus Stud 51:389–413

H&M Group. (2024). Sustainability. Retrieved from https://hmgroup.com/sustainability/. Accessed on 14 May 2025

Jamali D, Makarem Y, Willi A (2020) From diffusion to translation: implementation of CSR practices in MNC subsidiaries. Soc Respons J 16(3):309–327

Jolaoso S (2025) Oil clean-up 'scam' warnings ignored by Shell, whistleblower tells BBC. BBC. Retrieved from https://www.bbc.com/news/articles/c0rqe85q1jno. Accessed on 14 May 2025

Kolk A (2016) The social responsibility of international business: from ethics and the environment to CSR and sustainable development. J World Bus 51(1):23–34

Kuckartz U, Rädiker S (2019) Analyzing qualitative data with MAXQDA. Springer, Wiesbaden

Marschan-Piekkari R, Welch C (2011) Rethinking the case study in international business and management research. Edward Elgar Publishing, Cheltenham

Meyer JW, Rowan B (1977) Institutionalized organizations: formal structure as myth and ceremony. Am J Sociol 83(2):340–363

Miles H, Huberman AM, Saldana J (2014) Qualitative data analysis: a methods sourcebook. Sage, New York

Neumayer E, Perkins R (2005) Uneven geographies of organizational practice: explaining the cross-national transfer and diffusion of ISO 9000. Econ Geogr 81(3):237–259

Park BI, Ghauri PN (2015) Determinants influencing CSR practices in small and medium sized MNE subsidiaries: a stakeholder perspective. J World Bus 50(1):192–204

Preuss L, Barkemeyer R, Arora B, Banerjee S (2024) Sensemaking along global supply chains: implications for the ability of the MNE to manage sustainability challenges. J Int Bus Stud 55(4):492–514

Shein. (2025). Our Impact. Retrieved from https://www.sheingroup.com/our-impact/. Accessed on 14 May 2025

Simon H (2009) Hidden champions of the twenty-first century: success strategies of unknown world market leaders. Springer Science, New York

Swissinfo. (2020) Nestlé accused of making 'zero progress' on plastic pollution. Swissinfo. Retrieved from https://www.swissinfo.ch/eng/business/nestl%C3%A9-accused-of-making-zero-progress-on-plastic-pollution/46198410. Accessed on 14 May 2025

The Federal Government of Germany (2024) Greater protection for people and the environment. Retrieved from https://www.bundesregierung.de/breg-en/news/supply-chain-act-2250930. Accessed on 14 May 2025

Wehling C, Guanipa Hernandez A, Osland J, Osland A, Deller J, Tanure B,... Sairaj A (2009) An exploratory study of the role of HRM and the transfer of German MNC sustainability values to Brazil. Euro J Int Manag 3(2):176–198

Yin RK (2017) Case study research and applications: design and methods. Sage, Thousand Oaks

Yin J, Jamali D (2016) Strategic corporate social responsibility of multinational companies subsidiaries in emerging markets: evidence from China. Long Range Plan 49(5):541–558

Sustainability-Driven Solutions Can Generate Business Growth—A Real-Life Example by HP Inc.

Hristina Stoyanova and Atanas Lipov

Abstract HP Inc., a global leader in personal computer (PC) and printer manufacturing, in 2020 initiated a sustainability-driven initiative to extend the lifecycle of its PC devices, resulting in a significant reduction of CO_2e emissions from manufacturing. By refurbishing key external parts and cleaning functional internal components, HP successfully reduced the number of scrapped devices and lowered its carbon footprint. This approach not only proved to be environmentally beneficial but also commercially viable, leading to the creation of new service offerings for strategic commercial customers and the establishment of a dedicated "HP Renew Solutions" division. HP's sustainability efforts were recognized globally, with the company winning a TSIA Star Award for innovation in services in 2023. This business case explores HP's initiatives to decarbonize its IT operations and offer new, lower-carbon solutions to its strategic customers. The case also highlights the challenges faced, lessons learnt, benefits for customers and the business, and HP's goals for sustainability. It aims to contribute to the study of corporate sustainability practices, which are an essential tool in mitigating climate change.

Keywords Corporate sustainability · IT operations · Circular IT · Greenhouse gas emissions · Sustainable business practices

1 Introduction

Corporate sustainability has emerged as a relatively recent area of academic inquiry (Sabirali and Mahalakshmi 2023). Yet, it has rapidly gained traction among scholars (Mariani et al. 2021) due to the growing urgency to address the environmental impacts

H. Stoyanova (✉)
Sofia University St. Kliment Ohridski, Sofia, Bulgaria
e-mail: hristinais@uni-sofia.bg

A. Lipov (✉)
HP Inc., Sant Cugat del Valles, Spain
e-mail: nasko.lipov@gmail.com

A. Alvarez-Risco et al. (eds.), *Green International Business*, Environmental Footprints and Eco-design of Products and Processes,
https://doi.org/10.1007/978-3-032-12497-5_6

of climate change (Pillai et al. 2017). During the 2015 United Nations General Assembly, 17 Sustainable Development Goals (SDGs) were adopted, forming the foundation of the post-2015 Development Agenda (Spencer 2021). Among these, SDGs 9 and 12 highlight the critical role of business in addressing climate change (Nicolò et al. 2023). Corporations play a pivotal part in achieving sustainable development, particularly through innovation and the integration of sustainability into their core missions (Gallego-Sosa et al. 2020; Bastons et al. 2020). Given their significant role in resource consumption and economic activity, corporations are increasingly expected to lead sustainability efforts and serve as catalysts for sustainable development (Hart and Milstein 1999).

Despite the growing emphasis on sustainability, many companies remain convinced that increasing their environmental friendliness will erode their competitiveness. They believe that such efforts will add to costs and fail to deliver immediate financial benefits (Nidumolu et al. 2009). By examining HP Inc.'s initiative to decarbonize its IT operations, we argue that sustainability can be a key driver of innovation and new business growth.

2 About the Company

HP Inc. is a global personal computer (PC) and printer manufacturing company with a workforce of 57,000 employees. It operates in over 170 countries, consistently holding around 20% of the global PC market share alongside competitors such as Lenovo and Dell. The Information and Communication Technology (ICT) sector contributes significantly to global greenhouse gas emissions, estimated to account for 1.8 to 2.8% of total emissions in 2020 (Freitag et al. 2021). As a major player in this industry, HP contributes to these emissions through the production and distribution of IT hardware. In response to the increasing importance of sustainability in the global business landscape, HP has made environmental sustainability a core focus of its strategic agenda.

The HP Sustainability Policy emphasizes the company's commitment to environmental stewardship and social responsibility. It outlines strategic initiatives aimed at reducing greenhouse gas emissions, promoting circular economy practices, and enhancing digital equity. The company aims to achieve net-zero emissions by 2040 and reduce greenhouse gas (GHG) emissions by 50% by 2030 compared to 2019 levels. In 2023, HP's carbon footprint was 19,764,400 tons of CO_2e, representing a 27% decrease from 2019. Additionally, HP aims to use 100% renewable electricity in its operations by 2025, having already procured and generated 293,289 MWh of renewable electricity, equivalent to 59% of its global electricity consumption in 2023. The company also focuses on circularity, reaching 40% circularity in products and packaging by 2023, with a goal of 75% by 2030 (HP Inc. 2024). The company was ranked twentieth on TIME and Statista 2024 list of the world's most sustainable companies (TIME and Statista 2024).

3 The Goal

In 2020, the company started the transformation of its internal PC support model for all HP employees globally. One of the questions the team asked themselves was what they could do to eliminate or reduce the manufacturing CO_2 emissions while transforming their internal model. An option they identified was to extend the lifecycle of their PC devices.

4 The Solution

The company developed a new service model for its employees' business notebooks, which, through the refurbishment of key external parts and the clean-up of functional internal components, enabled the effective use of PCs for several additional years. As a result, HP scrapped 10,600 fewer devices during the period 2020–2023 and registered a significant reduction in CO_2e emissions from its internal operations. Learning from the car industry, which has predefined yearly maintenance schedules for its vehicles, the HP team also realized that it could leverage the proactive monitoring data collected from its devices in operation to make better and more sustainable decisions on which devices to refurbish proactively and which components to exchange. This allowed for a further reduction of the company's CO_2e emissions by a total of 456 tons per year for the internal HP services footprint alone. It was somewhat surprising to find that the main challenges with this new sustainability-driven approach were not coming from the outside market, but rather from within the company. The HP management had to navigate a mindset change from perceiving refurbished products as "second-hand" to being proud to support the global sustainability strategy.

5 The Commercial Adoption

After proving their business case internally, HP decided to incubate several service offerings in the market, which would enable strategic commercial customers to partner with HP and extend the lifecycle of selected devices within their IT environment. According to the company's estimates, this could unlock significant CO_2e emission savings of up to 20 to 40% for the entire IT fleet of customers, depending on the specifics of their IT environment.

6 The Business Challenge

While environmental sustainability is widely recognized as a strategic priority, the primary business challenge is the general perception that producing lower-carbon products and services always requires additional investment and would make them less competitive on the market compared to similar products or services that do not focus on sustainability benefits. HP was determined to prove to its customers that it was not true. The company decided to complement its portfolio instead of competing against its core offerings, and drive customer value in different ways. It offered its customers the ability to extend the lifecycle of their devices, enabling total cost of ownership (TCO) savings across their entire IT operations, while maintaining a healthy mix of new and refurbished devices. Thus, HP sought to identify new markets and revenue opportunities for its innovative, lower-carbon solutions.

7 The Result

Amid the COVID-19 pandemic in 2021, it became apparent that component and new device shortages presented a challenge. However, the emergence of this new "as-a-service" device life extension model offered a feasible alternative with incremental revenue potential, opening a new total addressable market (TAM) for companies focused on sustainability benefits and even monetizing them. As a result, HP expanded the model further with the implementation of a dedicated corporate "HP Renew Solutions" division, focused on refurbished products and services, and won a TSIA Star Award for innovation in services in 2023. The journey to more sustainable practices also helped HP implement other waste-reduction practices, such as the HOPE Recycling Futures (TM) program, which enables PC donations to underprivileged children, rather than recycling them.

8 Lessons Learned

A critical insight from HP's sustainability journey is that advancing environmental responsibility does not inherently require greater investment or lead to increased product and service costs. Instead, when environmental sustainability is strategically positioned as a core business driver, it enables companies to uncover new market opportunities, create innovative product and service offerings, and meet emerging customer demands. By integrating sustainability into its business model, HP demonstrated that lower-carbon solutions can not only be commercially viable but can also generate incremental profitability by addressing evolving environmental business needs.

9 Conclusion

HP's sustainability-driven initiative to extend the lifecycle of its PC devices exemplifies the integration of corporate sustainability practices into business strategies. As highlighted in the literature review, corporations play a pivotal role in achieving sustainable development through innovation and the incorporation of sustainability into their core missions (Gallego-Sosa et al. 2020; Bastons et al. 2020). HP's approach to refurbishing key external parts and cleaning functional internal components not only reduced CO_2e emissions but also proved commercially viable, aligning with the findings that sustainability efforts can lead to new market opportunities and innovative product offerings. The company's commitment to environmental stewardship and social responsibility, as outlined in the HP Sustainability Policy, demonstrates the growing emphasis on sustainability in the business landscape (Nicolò et al. 2023). By extending the lifecycle of their devices and implementing circular IT practices, HP successfully reduced its carbon footprint and developed new service offerings for strategic commercial customers. This initiative not only contributed to reducing greenhouse gas emissions but also enhanced HP's corporate reputation and ensured long-term success and resilience in a rapidly changing world.

HP's sustainability efforts underscore the importance of integrating environmental sustainability into business strategies, as they not only address the environmental impacts of climate change but also drive revenue growth. As the business landscape continues to evolve, the commitment to sustainability will remain a critical priority for corporations, enabling them to uncover new market opportunities and meet emerging customer demands.

References

Bastons M, Benguría R, Armengou J, Rey C (2020) Mission-based corporate sustainability: the Aigües de Barcelona model. In: Díaz BD, Capaldi N, Idowu SO, Schmidpeter R (eds) Responsible business in a changing world. Springer, pp 311–328. https://doi.org/10.1007/978-3-030-36970-5

Freitag C, Berners-Lee M, Widdicks K, Knowles B, Blair G, Friday A (2021) The real climate and transformative impact of ICT: a critique of estimates, trends, and regulations. Patterns 2(9):100340. https://doi.org/10.1016/j.patter.2021.100340

Gallego-Sosa C, Fernnndez-Torres Y, Gutierrez-Fernandez M (2020) Does gender diversity affect the environmental performance of banks? Sustainability 12(23). https://doi.org/10.3390/su12231 0172

Hart SL, Milstein MB (1999) Global sustainability and the creative destruction of industries. Sloan Manag Rev 41(1):23–33

HP Inc. (2024) Sustainable impact report executive summary 2023. Retrieved from https://h20195.www2.hp.com/v2/getpdf.aspx/c05179523.pdf

Mariani MM, Al-Sultan K, De Massis A (2021) Corporate social responsibility in family firms: a systematic literature review. J Small Bus Manag 1–55. https://doi.org/10.1080/00472778.2021.1955122

Nicolò G, Zanellato G, Tiron-Tudor A, Polcini PT (2023) Revealing the corporate contribution to sustainable development goals through integrated reporting: a worldwide perspective. Soc Respons J 19(5):829–857. https://doi.org/10.1108/SRJ-09-2021-0373

Nidumolu R, Prahalad CK, Rangaswami MR (2009) Why sustainability is now the key driver of innovation. Harv Bus Rev 87(9):57–64

Pillai KV, Slutsky P, Wolf K, Duthler G, Stever I (2017) Companies' accountability in sustainability: a comparative analysis of SDGs in five countries. In: Servaes J (ed) Sustainability development goals in the Asian context. Springer. https://doi.org/10.1007/978-981-10-2815-1

Sabirali KP, Mahalakshmi S (2023) Corporate sustainability practices: a systematic literature review and bibliometric analysis. Vis J Bus Perspecti. https://doi.org/10.1177/09722629231203125

Spencer J (2021) The sustainable development goals. In: Tsekleves E, Cooper R, Spencer J (eds) Design for global challenges and goals (1st edition, pp. 12–25). Routledge. https://www.taylorfrancis.com/chapters/edit/https://doi.org/10.4324/9781003099680-3/sustainable-development-goals-jak-spencer

TIME and Statista (2024) World's most sustainable companies 2024. TIME. https://time.com/collection/worlds-most-sustainable-companies-2024/

Beyond Recycling: How the Circular Economy is Reshaping Business

Katyuska Barja-Paredes

Abstract The linear economic model ("take-make-dispose") has intensified the climate crisis and waste generation, driven by increasing consumption and designs that ignore end-of-life considerations. In the face of this, the circular economy is emerging as a transformative paradigm that prioritises eliminating waste by design, maintaining products and materials at their maximum utility, and regenerating natural systems. Foundational concepts such as the closed systems vision, technical-biological cycles, and ecological-social limits underscore the need to transition towards regenerative models. Leading companies demonstrate their viability through strategies such as furniture refurbishment, repair/buyback programmes, and service (rather than sales) models. This generates key economic benefits: cost reduction through material efficiency, business model innovation, resilience to resource scarcity, and significant emissions reductions. The transition faces barriers such as fragmented regulations, complex supply chains, and linear consumption habits, but is driven by critical enablers, including digital technologies (IoT, AI), supportive policies (e.g. the right to repair), and multisector collaboration. Their accelerated adoption is crucial for achieving global climate goals and building inclusive, resilient economies.

Keywords Circular economy · Circularity · Circular design · Circular business · Sustainability

K. Barja-Paredes (✉)
University of York, York, England
e-mail: kbarjap@gmail.com

A. Alvarez-Risco et al. (eds.), *Green International Business*, Environmental Footprints and Eco-design of Products and Processes,
https://doi.org/10.1007/978-3-032-12497-5_7

1 Moving Past Recycling Towards Circularity

The unprecedented growth of the global population, coupled with rising standards of living, has precipitated a dramatic increase in the consumption of goods and energy. While this trend reflects economic development, it has also intensified environmental pressures concurrently. Elevated consumption patterns are strongly correlated with increased waste generation and higher energy use, both of which significantly contribute to climate change (Malinauskaite et al. 2017). Central to this issue is the prevailing linear model of industrialisation—often summarised as "take-make-dispose"—which has rendered waste management a defining challenge of the twenty-first century (Minelgaitė and Liobikienė 2019).

To effectively mitigate the escalating consequences of climate change, a paradigm shift is required. Strategies must prioritise resource efficiency, foster transitions to clean energy, and support the development of sustainable mobility systems. Of particular concern is the unsustainable consumption of raw materials. Though foundational to the global economy, the extraction and processing of these materials have been major contributors to greenhouse gas emissions over the past several decades.

Contrary to the popular belief that waste is an inevitable outcome of modern life, it is more accurately understood as a by-product of deliberate design choices. Natural ecosystems demonstrate the feasibility of waste-free systems: every output functions as an input within a regenerative cycle. In contrast, anthropogenic systems frequently neglect end-of-life planning during the design phase, whether for disposable items like snack wrappers or durable structures such as buildings and infrastructure (Ellen MacArthur Foundation 2022b).

The Intergovernmental Panel on Climate Change (IPCC 2023) has documented the economic repercussions of anthropogenic climate change across multiple sectors, including agriculture, energy, fisheries, forestry, and tourism. Similarly, the Global Risks Report (World Economic Forum 2025) identifies "extreme weather events" as the most probable environmental threat capable of inciting large-scale global crises. Recent episodes, ranging from heat waves in Asia to wildfires in North America, underscore the urgent need for systemic interventions. In response, the international community has committed, through the Paris Agreement, to limit the global average temperature increase to well below 2 °C above pre-industrial levels, with efforts to cap the rise at 1.5 °C (United Nations 2015). Meeting these targets demands not only increased climate resilience but also the structural transformation of economic systems, from a linear to a circular model (Skinmo and Fjærli 2021).

Recycling, although important, is no longer sufficient. It presents inherent limitations: many materials are either non-recyclable or suffer from quality degradation through repeated cycles (Ohiomah and Sukdeo 2022). Additionally, some recycling processes are more energy-intensive than producing virgin materials. As such, recycling should be viewed as a single step within a broader hierarchy of resource strategies that also includes reduction, reuse, remanufacture, and recovery (Allwood

2014). Compared to reuse and remanufacturing, recycling often captures less material value, and scholars caution against mischaracterising recycling as a closed-loop system (Braungart and McDonough 2009; Mishra et al. 2018).

In this broader context, the circular economy emerges as a transformative alternative. It rejects the linear conception of waste and seeks to retain materials, products, and components in circulation at their highest utility and value, for as long as possible, with minimal system leakages (Korra 2022; Spring and Araujo 2017). Environmentally, the circular economy reduces emissions, conserves resources, and relieves ecological stress (Tunn et al. 2019). Economically, it offers advantages including reduced material costs, greater resilience to price volatility, and enhanced supply chain robustness (Ellen MacArthur Foundation et al. 2015; Shpak et al. 2020).

While implementation will vary across geographies and sectors, the transition from linear to circular models is increasingly recognised as a critical pathway towards sustainable, inclusive, and resilient development.

2 Foundations and Evolution of the Circular Economy Concept

Advanced circular economy (CE) systems and sophisticated recycling technologies form the cornerstone of a resource-efficient, sustainable society (Hagelüken and Goldmann 2022). The CE can be understood as the integration of economic activity and environmental well-being (Murray et al. 2017). The CE model emerges not merely as a conceptual counterpoint but as a viable and urgently needed alternative to the dominant linear economy, which is primarily concerned with wealth creation through a linear chain of extraction, production, distribution, consumption, and disposal (Sariatli 2017). This extractive model has led to severe ecological degradation and social inequalities, as recognised by numerous global institutions (European Commission 2024; Intergovernmental Panel on Climate Change 2023; World Economic Forum 2024). Over time, the concept of the circular economy has evolved to incorporate diverse frameworks, expanding its definition to include more concrete, measurable, and implementable principles. This section outlines five key conceptual developments that have shaped the CE discourse, reflecting its growing maturity and practical relevance in addressing global sustainability challenges.

2.1 Kenneth Boulding's "Spaceman Economy"

One of the earliest conceptual foundations for the concept of the CE was proposed by economist Kenneth Boulding in his seminal 1966 essay, "The Economics of the Coming Spaceship Earth" (Wautelet 2018). Boulding distinguishes between two contrasting economic paradigms: the "cowboy economy" and the "spaceman economy." The former mirrors the frontier mindset of the nineteenth century American West—an open system characterised by unlimited resource exploitation and waste generation. In contrast, the "spaceman economy" envisions Earth as a closed system, similar to a spacecraft, where finite resources must be carefully managed and waste efficiently repurposed. In this vision, economic activities are inherently interlinked with environmental and social systems, necessitating an integrated, circular approach to sustainability. Boulding's work laid the theoretical groundwork for recognising ecological limits within economic models (Boulding 1966; Wautelet 2018).

2.2 Cradle-to-Cradle Thinking

In 2002, architect William McDonough and chemist Michael Braungart introduced the "Cradle-to-Cradle" design framework in their influential publication *Cradle to Cradle: Remaking the Way We Make Things* (Wautelet 2018). This model draws a sharp contrast between the biological, regenerative cycles observed in nature and the linear, extractive processes of modern industry, referred to as "cradle-to-grave." The authors propose that products should be designed from the outset with their end-of-life in mind, ensuring that materials can either be safely returned to the environment or reintegrated into production cycles. This philosophy, which considers waste as "food" for future cycles, encourages the design of products and systems that support long-term environmental balance and material conservation (Braungart and McDonough 2009).

2.3 The Ellen MacArthur Foundation and the Butterfly Diagram

The establishment of the Ellen MacArthur Foundation in 2010 significantly accelerated the mainstreaming of circular economy principles among business, academic, and policy communities. In their foundational 2013 report *Towards the Circular Economy: Economic and Business Rationale for an Accelerated Transition*, the Foundation articulated three core principles central to CE (Ellen MacArthur Foundation 2013):

i. **Design out waste and pollution** by rethinking product and system design.
ii. **Keep products and materials in use**, ensuring their highest value is retained over time.
iii. **Regenerate natural systems**, restoring ecological health rather than depleting it.

The report also introduced the now-iconic *Butterfly Diagram*, a visual representation of the two key cycles in CE: the *technical cycle* (right wing), which involves reuse, repair, remanufacture, and recycling of products and components, and the *biological cycle* (left wing), which focuses on returning organic materials safely to the biosphere. This diagram helped translate CE theory into actionable strategies for design and production systems.

2.4 Kate Raworth's Doughnut Economics

In 2017, economist Kate Raworth introduced the "Doughnut Economics" framework, which complements and extends the principles of CE through a socio-ecological lens. The doughnut diagram consists of two concentric circles: an "inner social foundation" that ensures universal access to life's essentials, and an "outer ecological ceiling" that defines planetary boundaries (Raworth 2017). The safe and just space between these rings, resembling a doughnut, is where humanity can thrive sustainably.

Raworth proposes seven principles for 21st-century economics, including the imperative to "create to regenerate.“ She aligns with the Ellen MacArthur Foundation's biological cycle but critiques the technical cycle, arguing that no industrial system can achieve perfect material recovery. Hence, she suggests replacing the term "circular economy" with "cyclical economy" to reflect the limitations inherent in technical recovery loops. Her framework encourages businesses and governments to shift their focus from pursuing infinite growth to embracing a regenerative, distributive economy that respects both ecological limits and social needs.

2.5 A Scientific Definition: Nobre and Tavares (2021)

In an effort to consolidate the many interpretations of the CE, Nobre and Tavares (2021) conducted an extensive literature review and 44 expert interviews to formulate a scientifically grounded and operational definition. They define the circular economy as:

> "An economic system that targets zero waste and pollution throughout materials' lifecycles, from environmental extraction to industrial transformation and final consumption, applying to all involved ecosystems. Upon its lifetime end, materials return to either an industrial process or, in the case of treated organic residuals, safely back to the environment, completing a natural regenerating cycle. It operates by creating value at the macro, meso and micro levels and exploits to the fullest the sustainability nested concept. Used energy sources are clean and renewable. Resource use and consumption are efficient. Government agencies and responsible consumers play an active role in ensuring the correct system's long-term operation." (Nobre and Tavares 2021, p.10). (Nobre and Tavares 2021)

This comprehensive definition captures the multi-scalar, systemic nature of CE and incorporates key sustainability principles, from clean energy use to stakeholder engagement. These evolving conceptualisations reflect the increasing maturity and adaptability of the circular economy paradigm. Each framework contributes to a more holistic understanding of how economic systems can be designed to operate within environmental boundaries while ensuring long-term societal well-being.

3 Business Transformation in Action

The transition towards a circular economy necessitates that businesses fundamentally rethink how value is created, delivered, and preserved throughout the product life cycle (Kalmykova et al. 2018). Rather than treating circularity as an environmental add-on, leading companies should embed it into their core strategies, supply chains, and customer relationships as a new business model (Ellen MacArthur Foundation 2013). This transformation is manifesting in a variety of innovative models, including product-as-a-service systems (Matschewsky 2019; Pieroni et al. 2019; Vezzoli et al. 2021), remanufacturing (Hanes 2018; Ohiomah and Sukdeo 2022), and repair schemes (Bocken et al. 2016; Stahel 2019), among others. This section explores how pioneering firms, such as IKEA, Patagonia, and Philips, are operationalizing circular economy principles to not only reduce environmental impact but also unlock new forms of long-term value creation.

3.1 IKEA: Furniture Take-Back, Resale, and Circular Design

IKEA, one of the world's largest furniture retailers, has committed to transitioning towards a circular business and using only renewable or recycled materials by 2030 (IKEA 2025b). Recognising the environmental footprint of its mass-produced furniture, the company has launched several initiatives aimed at closing material loops and extending product lifespans, as outlined in its latest report, the "IKEA Circular Product Design Guide" of 2024. One of IKEA's key strategies is the development of a furniture take-back and resale programme, piloted in Sweden and now available in all their stores. In the programme "Buyback and Resell" service, customers can return used furniture, which is then assessed, remanufactured, and resold as a second-hand item in-store (IKEA 2025a, 2025c). This initiative not only reduces waste but also promotes a cultural shift towards reuse and refurbishment. In addition, IKEA is investing heavily in circular product design, aiming to ensure that all products are made from renewable or recycled materials and are designed for easy disassembly and repair (IKEA 2024, 2025b). This strategic integration of circularity at the design stage is crucial for enabling scalable, post-consumer recovery systems.

3.2 *Patagonia: The Worn Wear Programme and Durable Design*

Patagonia, the outdoor apparel company, is a longstanding advocate of environmental stewardship and has embraced circular economy principles through its "Worn Wear" programme (Archana 2021). Launched in 2012, this initiative encourages customers to repair, reuse, and trade in their used clothing, which is then cleaned, refurbished, and resold at reduced prices (Patagonia 2025; Patagonia UK 2025). By promoting a "buy less, repair more" philosophy, Patagonia has deliberately challenged the fast fashion model, demonstrating that profitability and sustainability can be mutually inclusive (Chouinard 2023; Rattalino 2018). In parallel, the company designs its products for durability and repairability, offering online tutorials, repair guides, and, if customers do not want to make the repairs themselves, (Patagonia UK 2025). Patagonia's approach aligns with the circular economy principle of extending product life and reducing material through high quality and durable products (Bocken et al. 2016). The company also measures and reports on the carbon and water savings associated with its reuse activities, helping to embed transparency and accountability into its circular strategy.

3.3 *Philips: Product-as-a-Service and Industrial Circularity*

Philips, a global leader in health technology and lighting solutions, has embraced circular economy innovation through its Product-as-a-Service model. Rather than selling products, Philips retains ownership of the equipment and instead offers lighting as a subscription service: managing installation, maintenance, and upgrades throughout the product's life cycle (Philips 2025a). This model incentivizes design for longevity, modularity, and easy disassembly, as Philips remains responsible for the system's performance and end-of-life recovery (Bressanelli et al. 2019). The PaaS model has been implemented in several high-profile projects, including Schiphol Airport in the Netherlands, where Philips provides a long-term lighting solution while maintaining full material stewardship. They also have another similar project in Argentina and Singapore (Philips 2017; Van Houten 2014). Beyond lighting, Philips has committed to embedding circular practices across its value chain. The company aims to generate 25% of its revenue from circular products and services and is actively pursuing remanufacturing programmes for medical imaging equipment (Philips 2025b). These remanufactured systems not only extend the useful life of complex technology but also reduce waste and resource use without compromising quality or performance.

4 Economic and Environmental Benefits of the Circular Economy

The circular economy is often viewed through an environmental lens, but its value extends far beyond ecological gains (Ellen MacArthur Foundation et al. 2015; Hagelüken and Goldmann 2022; Korra 2022; Shpak et al. 2020; Spring and Araujo 2017; Tunn et al. 2019). By fundamentally rethinking how resources are used, recovered, and circulated, the circular economy offers a compelling array of economic advantages that can enhance business performance, foster innovation, and strengthen long-term competitiveness (Ellen MacArthur Foundation et al. 2015; Bressanelli et al. 2019; Jabbour et al. 2019; Da Costa Fernandes and Rozenfeld 2024). Additionally, it supports the achievement of international climate targets and aligns closely with Environmental, Social, and Governance (ESG) frameworks, which are increasingly demanded by investors, regulators, and consumers (Mascotto 2020; Provensi et al. 2025). This section outlines the multifaceted benefits of circularity, focussing on cost efficiency, innovation, competitive positioning, and climate responsibility.

4.1 *Cost Savings Through Resource Efficiency*

One of the most immediate and tangible benefits of adopting circular economy practices lies in resource efficiency. By designing out waste, optimising materials use, and improving energy performance, businesses can significantly reduce their operational costs (Ellen MacArthur Foundation 2013; Ellen MacArthur Foundation et al. 2015). The circular model challenges the prevailing "take-make-waste" logic and shifts attention towards retaining value for as long as possible. For example, remanufacturing and refurbishment can cost significantly less than producing new items from virgin resources, savings that are especially relevant in sectors with high material input costs, such as electronics, the automotive industry, and industrial machinery (Hagelüken and Goldmann 2022). According to the Ellen MacArthur Foundation (2013), implementing circular economy principles in the EU could generate net material cost savings of up to USD 600 billion annually by 2025, particularly in medium-lifetime, complex products such as motor vehicles. Furthermore, circular supply chains are more resilient to raw material volatility and price shocks (Ellen MacArthur Foundation 2013; Lewandowski 2016; Kalmykova et al. 2018). As seen during the COVID-19 pandemic and subsequent geopolitical disruptions, global supply chains proved vulnerable to resource scarcity. Circular strategies, such as localised reuse and diversified material sourcing, can mitigate these risks, ensuring greater business continuity (Heshmati 2017).

4.2 *Innovation and New Business Models*

The shift towards circularity encourages systemic innovation, challenging firms to redesign products, services, and delivery mechanisms (World Economic Forum 2014). This has encouraged the development of new business models, including:

- Product-as-a-Service (PaaS): where ownership remains with the producer and customers pay for usage, e.g. Philips' lighting-as-a-service (Philips 2017; Pieroni et al. 2019; Matschewsky 2019; Vezzoli et al. 2021).
- Reverse logistics and recovery models: for repair, resale, or remanufacturing; e.g. Patagonia's Worn Wear (Bocken et al. 2016; Stahel 2019; Chouinard 2023).
- Sharing platforms: which enable better utilisation of idle assets; e.g. Airbnb, Zipcar (Schwanholz and Leipold 2020).

These models often foster stronger customer relationships, as businesses become responsible for the entire product life cycle. Moreover, innovation in circular design has led to material breakthroughs, such as compostable bioplastics (Baskoro et al. 2024) or textiles made from regenerated fibres (Kim et al. 2022). These advances not only address sustainability but also create differentiated product offerings in increasingly saturated markets (Carra and Magdani 2017).

4.3 *Competitive Advantage and Reputational Value*

Circularity is emerging as a key strategic differentiator in today's market landscape. Consumers, particularly younger generations, are more inclined to support brands that align with their environmental and social values (Nguyen et al. 2018; Uddin and Khan 2016). According to McKinsey and NielsenIQ (2023), consumers consider a sustainable lifestyle to be important, and they expect companies to take a proactive role in promoting it (McKinsey and NielsenIQ 2023). By adopting transparent circular practices, companies can enhance their brand equity and customer loyalty, gaining a reputational edge over competitors still anchored in linear models (Michelon 2011). Circularity can also open access to green financing, as investors increasingly favour companies with strong ESG credentials (Mascotto 2020; Provensi et al. 2025). Additionally, early adoption of circular principles enables companies to anticipate and comply with evolving regulatory frameworks, such as the European Green Deal, the EU Circular Economy Action Plan and extended producer responsibility (EPR) (Ellen MacArthur Foundation 2022a). By staying ahead of these policies, businesses reduce compliance risks and position themselves as leaders in the transition to sustainable markets (PwC 2025).

4.4 Contribution to Climate Goals and ESG Compliance

The circular economy plays a crucial role in helping nations and companies achieve their climate targets, as outlined in the Paris Agreement. By reducing the need for virgin resource extraction and energy-intensive production, circular strategies can significantly lower greenhouse gas (GHG) emissions. According to the Circularity Gap Report (2021), transitioning to a circular economy could cut global emissions by 39% and help avoid climate breakdown (Circle Economy 2021). From an ESG perspective, circularity supports environmental goals by reducing pollution and emissions, promotes social value through local job creation, and enhances governance by improving transparency in supply chains (Marcon et al. 2023). Recent research on firms in Sub-Saharan Africa (2010–2022) indicates that environmental and social ESG disclosures drive the adoption of circular economies, while weak governance frameworks can hinder progress (Agyemang et al. 2025). The presence of sustainability committees and investment in eco-innovation significantly strengthens this relationship, highlighting the need for robust ESG reporting, strategic leadership, and innovation to accelerate circular transitions, especially in emerging markets. Moreover, many circular economy practices are measurable and auditable, enabling organisations to report progress in line with ESG disclosures and align with international standards such as the Global Reporting Initiative (GRI). Demonstrating strong circular initiatives not only enhances transparency and accountability but can also positively influence a company's reputation and revenue performance (Panori et al. 2024).

5 Barriers and Enablers of the Circular Economy

While the circular economy (CE) presents a promising pathway towards sustainable development, its widespread implementation is not without challenges. Transitioning from a linear to a circular model requires systemic change across industries, governance structures, and consumer cultures (Cantú et al. 2021; Heshmati 2017; Ohiomah and Sukdeo 2022; Ormazabal et al. 2018; Rizos et al. 2016). This section explores the key barriers impeding progress and the critical enablers that can accelerate adoption at scale.

5.1 Barriers to Circular Economy Implementation

Despite growing interest and momentum, several persistent barriers hinder the effective implementation of circular strategies in business and society. These include:

(a) **Supply Chain and design complexity**

Modern supply chains are global, complex, and optimised for linear throughput, making circular implementation difficult. Their fragmented nature complicates material traceability, logistics, and end-of-life recovery, mainly when components are sourced from multiple countries and suppliers (Pan et al. 2015). One core challenge is that products are often not designed with circularity in mind, resulting in difficulties with disassembly, refurbishment, or remanufacturing. Supply chain partners must therefore coordinate from the design phase to standardise materials and processes that support reuse. However, many companies lack the technical capacity, skilled labour, and clear guidelines to implement such changes across supply networks. Without strong collaboration, transparency, and harmonised standards throughout the chain, circular supply models remain difficult to scale (Bressanelli et al. 2019; Govindan and Hasanagic 2018; Mishra et al. 2018).

(b) **Consumer Behaviour and Cultural Norms**

Consumer habits remain predominantly aligned with the linear economy, characterised by fast consumption and frequent disposal. Preferences for low-cost, disposable goods, limited awareness of sustainable alternatives, and the perceived inconvenience of reuse, repair, or return pose significant barriers to the adoption of circular business models, especially in high-consumption societies (Auerbach George et al. 2023). These behavioural challenges are compounded by deep-rooted cultural norms and values, which often resist shifts towards sustainable consumption. Camacho-Otero et al. (2018) discuss in their comprehensive literature review, which analyses 111 studies on circular economy consumption, that consumer acceptance remains a critical obstacle. While much of the research focuses on identifying drivers and barriers, emerging studies suggest that consumption in the circular economy is shaped by complex, multi-value systems, where choices are influenced not only by utility but also by identity, trust, and political values. Addressing these challenges requires more than awareness campaigns; it demands systemic education, design thinking that integrates consumer perspectives, and digital tools to support behaviour change and facilitate adoption at scale (Camacho-Otero et al. 2018).

(c) **Inadequate Infrastructure**

A significant barrier to circularity in low-income countries is the inadequate waste management infrastructure, particularly in rural areas where informal systems often dominate. Studies in Ghana demonstrate that while innovative technologies (e.g. sensor bins, GPS tracking) show promise, foundational infrastructure must be strengthened first to enable their effective use (Anokye et al. 2024). More broadly, municipal waste systems in developing countries face persistent challenges, including insufficient funding, rapid urbanisation, and heightened environmental risks (Malinauskaite et al. 2017). Although solutions like decentralised systems, stakeholder engagement, and the inclusion of informal recyclers present viable and sustainable pathways, these must be built on a base of robust infrastructure and governance. Overall, the findings highlight that while innovation can support circularity, the absence of enabling infrastructure remains a critical obstacle in low-income contexts.

(d) **Policy Gaps and Inconsistent Regulations**

Although sustainability has become a political priority in regions like the EU, policy frameworks supporting the circular economy remain inconsistent and fragmented. In some jurisdictions, outdated regulations, such as restrictive classification standards or tax regimes, actively disincentivise reuse, remanufacturing, and renewable integration. For multinational companies, a lack of harmonised policies across borders complicates implementation. Carra and Magdani (2017) highlight that government policy can either facilitate or hinder the shift towards a circular economy. Key areas such as consumption taxes, legal frameworks, industrial strategies, and building regulations require reform to support circular practices. For instance, in the UK, new construction is tax-exempt, whereas refurbishment is subject to a 20% VAT, potentially deterring retrofit initiatives. In contrast, some European countries are beginning to adjust their tax systems. Sweden, for example, has reduced taxes on repair services to encourage more sustainable consumption (Carra and Magdani 2017).

5.2 *Enablers Driving Circular Transformation*

In contrast, several powerful enablers are emerging that can help businesses, policymakers, and civil society overcome these challenges and unlock the full potential of the circular economy:

(a) **Digital Technologies: IoT, Blockchain, and AI**

The rise of smart manufacturing is increasingly driven by technological innovations, such as the Internet of Things (IoT), which enables machines and systems to communicate in real-time, allowing for more efficient monitoring, control, and optimization of industrial processes. Industrial IoT connects devices, operators, and management through high-speed data flows, enhancing productivity and operational transparency. These advancements are pivotal for circularity: IoT enables real-time tracking of product performance and location, facilitating timely maintenance, reuse, and refurbishment (Ramadoss et al. 2018). Complementing this, blockchain technology enables circular sourcing of renewable inputs and supports resource efficiency. It can also aid in the recovery of materials, particularly through refurbishing and recycling by manufacturers and consumers, by tracking material and resource flows through various supply chains and consumption steps (Böckel et al. 2021). Artificial Intelligence (AI) plays a pivotal role in advancing circular business models by modelling product lifecycles, predicting maintenance needs, and optimising material flows. It enables practical applications such as dynamic pricing to reduce food waste by lowering prices as products near expiry, matching algorithms that efficiently connect users with second-hand goods or shared resources, and predictive demand systems that enhance reverse logistics for items like glass or medicines (Montes-Pineda and Garrido-Yserte 2024). Rather than simply reviewing past implementations, current research highlights how firms can proactively leverage AI to embed circularity into operations, transforming how products are designed, used, and recovered. Together, these technologies provide powerful tools to integrate circular principles into industrial operations, transforming the way products are designed, used, and recovered.

(b) **Supportive Regulatory Frameworks**

Although it is well known that policy gaps hinder the implementation of the circular economy, some progressive legislative tools and frameworks are beginning to shape a more circular approach, such as:

- The "European Green Deal," which includes the "Circular Economy Action Plan," aims to decouple economic growth from resource use by promoting sustainable product design, empowering consumers, and establishing right-to-repair rules (Ellen MacArthur Foundation 2022a).
- The "Right to Repair" legislation, implemented in various jurisdictions including the EU, mandates manufacturers to make spare parts and repair information accessible, extending product lifespans (European Parliament 2024).
- Extended Producer Responsibility (EPR) schemes are holding producers accountable for the end-of-life management of their products, incentivising circular design and reducing landfill reliance (PwC 2021).

These policies not only mandate change but also level the playing field, allowing circular innovators to compete fairly with traditional producers.

(c) **Collaborative Platforms and Ecosystems**

Collaborative platforms and ecosystems are vital to advancing the circular economy, as systemic change cannot be achieved by individual actors alone. Industry coalitions like the Circular Economy 100 (CE100) by the Ellen MacArthur Foundation promote cross-sector knowledge-sharing and co-innovation (Ellen MacArthur Foundation 2025), while regional initiatives such as Circular Glasgow (Glasgow Chamber of Commerce 2021) and Amsterdam's Doughnut Coalition (Doughnut Economics Action Lab 2020) model demonstrate how circular practices can be successfully piloted at the city level and scaled more broadly. In Latin America and the Caribbean, the Circular Economy Coalition, supported by the European Union, was launched in 2021, bringing together government agencies, the private sector, academia, and international partners in a multi-stakeholder effort to promote circularity (United Nations Environment Programme 2021). These ecosystems play a critical role in establishing shared standards, aligning goals, and coordinating actions, thereby enhancing efficiency and maximising collective impact.

6 Conclusion

The circular economy is no longer a fringe concept or a temporary sustainability trend; it is rapidly becoming a strategic imperative for long-term business resilience, economic prosperity, and environmental stewardship. In the face of escalating global challenges, climate change, biodiversity loss, resource depletion, and mounting waste, linear models of production and consumption are proving increasingly obsolete and unsustainable. Circularity offers not only a solution but a vision: one where growth is decoupled from resource extraction, where innovation drives regeneration, and where businesses thrive by creating value that lasts.

For companies across all sectors and geographies, transitioning to circular models represents both a responsibility and an opportunity. Those who lead this transformation are better positioned to future-proof their operations, access new markets, comply with increasingly stringent environmental regulations, and meet the rising expectations of investors, consumers, and employees. Far from being a constraint, circularity unlocks a new wave of creativity, efficiency, and collaboration that will define the next era of industry and commerce. However, this shift demands more than incremental improvements. It calls for systemic thinking, cross-sector collaboration, and bold leadership. Business leaders must reimagine their value chains, from design to end-of-life, and integrate regenerative principles into their corporate strategy. They must also work closely with policymakers, innovators, and civil society to scale circular solutions globally. Education, digitalization, and transparent metrics will be crucial enablers in accelerating this shift.

As this chapter has illustrated, circularity touches every dimension of sustainable business, from operations and innovation to global trade and policy. It is both a practical framework and a moral compass for navigating a rapidly changing world. The businesses that embrace it today will not only lead the transition but also help shape a more inclusive, resilient, and regenerative global economy for generations to come.

References

Agyemang AO, Osei A, Kongkuah M (2025) Exploring the ESG-circular economy nexus in emerging markets: a systems perspective on governance, innovation, and sustainable business models. Bus Strateg Environ n/a(n/a). https://doi.org/10.1002/bse.4278

Allwood JM (2014) Chapter 30-squaring the circular economy: the role of recycling within a hierarchy of material management strategies. En: Worrell E., Reuter MA (eds) Handbook of recycling. Elsevier, pp 445–477. https://doi.org/10.1016/B978-0-12-396459-5.00030-1

Anokye K, Apea OB, Darko AO, Kwabi LA (2024) The future of waste management in Ghana: assessing the feasibility and scalability of smart waste solutions amidst key adoption challenges. Cleaner Waste Syst 9:100183. https://doi.org/10.1016/j.clwas.2024.100183

Archana R (2021) Our quest for circularity. Patagonia. https://www.patagonia.com/stories/our-quest-for-circularity/story-96496.html

Auerbach George H, Tregenza L, Stenton M, Kapsali V, Blackburn RS, Houghton JA (2023) Challenging perceptions of fast and slow in contemporary fashion: a review of the paper dresses trend in the United Kingdom and the United States during the 1960s. Int J Sustain Fash Textil 2(1):Article 1. https://eprints.whiterose.ac.uk/id/eprint/198604/
Baskoro ML, Tjahjono B, Beltran M, Bogush A, Wang Y (2024) The imperative of communication signals in boosting business strategies of the bioplastic packaging industry. Bus Strateg Environ 33(2):Article 2. https://eprints.whiterose.ac.uk/id/eprint/201132/
Böckel A, Nuzum A-K, Weissbrod I (2021) Blockchain for the circular economy: analysis of the research-practice gap. Sustain Product Consumpt 25:525–539. https://doi.org/10.1016/j.spc.2020.12.006
Bocken NMP, de Pauw I, Bakker C, van der Grinten B (2016) Product design and business model strategies for a circular economy. J Ind Prod Eng 33(5):308–320. https://doi.org/10.1080/21681015.2016.1172124
Boulding KE (1966) The economics of the coming spaceship earth. https://arachnid.biosci.utexas.edu/courses/THOC/Readings/Boulding_SpaceshipEarth.pdf
Braungart M, McDonough W (2009) Cradle to cradle: Remaking the way we make things/William McDonough and Michael Braungart. Vintage Books. https://www.vlebooks.com/Product/Index/273455?page=0&startBookmarkId=-1
Bressanelli G, Perona M, Saccani N (2019) Challenges in supply chain redesign for the circular economy: a literature review and a multiple case study. Int J Prod Res 57(23):7395–7422. https://doi.org/10.1080/00207543.2018.1542176
Camacho-Otero J, Boks C, Pettersen IN (2018) Consumption in the circular economy: a literature review. Sustainability 10(8):Article 8. https://doi.org/10.3390/su10082758
Cantú A, Aguiñaga E, Scheel C (2021) Learning from failure and success: the challenges for circular economy implementation in SMEs in an emerging economy. Sustainability 13(3):Article 3. https://doi.org/10.3390/su13031529
Carra G, Magdani N (2017) Circular business model for the built environment. https://content.ellenmacarthurfoundation.org/m/17a5252e4dd10230/original/Circular-Business-Model-for-the-Built-Environment.pdf?_gl=1*109tjci*_gcl_au*MTQ3NDQ0MjcwNC4xNzQwMTQxODEw*_ga*NDU3MzcxMDkwLjE3MzE2OTAzOTE.*_ga_V32N675KJX*czE3NDc4OTc2ODgkbzgkZzEkdDE3NDc4OTk3OTYkajM0JGwwJGgwJGRvWm4yYm9xcFpaZ3gzZmNlR1FtSTR2NEFrTlAyME5MRm9R
Chouinard Y (2023) Patagonia shows how turning a profit doesn't have to cost the Earth (T. Hansen) [Entrevista]. https://www.mckinsey.com/industries/agriculture/our-insights/patagonia-shows-how-turning-a-profit-doesnt-have-to-cost-the-earth
Circle Economy (2021) Circularity Gap Report 2021. https://cdn.prod.website-files.com/5d26d80e8836af2d12ed1269/60210bc3227314e1d952c6da_20210122%20-%20CGR%20Global%202021%20-%20Report%20-%20210x297mm.pdf
Da Costa Fernandes S, Rozenfeld H (2024) Business model innovation through the design of circular product-service system value propositions: a method proposal. Bus Strateg Environ 33(6):5325–5345. https://doi.org/10.1002/bse.3759
Doughnut Economics Action Lab (2020) Amsterdam City Doughnut. https://doughnuteconomics.org/stories/amsterdam-city-doughnut
Ellen MacArthur Foundation (2013) Towards the circular economy Vol 1 an economic and business rationale for an accelerated transition.pdf. https://emf.thirdlight.com/file/24/xTyQj3oxiYNMO1xTFs9xT5LF3C/Towards%20the%20circular%20economy%20Vol%201%3A%20an%20economic%20and%20business%20rationale%20for%20an%20accelerated%20transition.pdf
Ellen MacArthur Foundation (2022a) The EU's circular economy action plan. https://www.ellenmacarthurfoundation.org/circular-examples/the-eus-circular-economy-action-plan
Ellen MacArthur Foundation (2022b) Eliminate waste and pollution. https://www.ellenmacarthurfoundation.org/eliminate-waste-and-pollution
Ellen MacArthur Foundation (2025) The world's leading circular economy network. https://www.ellenmacarthurfoundation.org/network/overview

Ellen MacArthur Foundation, Stiftungsfonds für Umweltökonomie und Nachhaltigkeit, and McKinsey Center for Business and Environment (2015) Growth within a circular economy vision for a competitive Europe. https://emf.thirdlight.com/file/24/_A-BkCs_h7gRYB_Am9L_JfbYWF/Growth%20within%3A%20a%20circular%20economy%20vision%20for%20a%20competitive%20Europe.pdf

European Commission (2024) Consequences of climate change. https://climate.ec.europa.eu/climate-change/consequences-climate-change_en

European Parliament (2024) Right to repair: making repair easier and more appealing to consumers. https://www.europarl.europa.eu/news/en/press-room/20240419IPR20590/right-to-repair-making-repair-easier-and-more-appealing-to-consumers

Glasgow Chamber of Commerce (2021) What is circular glasgow?–circular glasgow. https://www.circularglasgow.com/what-is-circular-glasgow/

Govindan K, Hasanagic M (2018) A systematic review on drivers, barriers, and practices towards circular economy: a supply chain perspective. Int J Prod Res 56(1/2):278–311. https://doi.org/10.1080/00207543.2017.1402141

Hagelüken C, Goldmann D (2022) Recycling and circular economy—towards a closed loop for metals in emerging clean technologies. Miner Econ 35(3):539–562. https://doi.org/10.1007/s13563-022-00319-1

Hanes R (with Nicholson S, Carpenter A, National Renewable Energy Laboratory) (2018) Evaluating opportunities for a circular economy: an investigation of economic and decision-making underlying recycling, reusing and remanufacturing habits in U.S. consumers and manufacturers. National Renewable Energy Laboratory

Heshmati A (2017) A review of the circular economy and its implementation. Int J Green Econ 11(3–4):251–288. https://doi.org/10.1504/IJGE.2017.089856

IKEA (2024) IKEA circular product design guide: guide to designing products with circular capabilities. https://www.ikea.com/global/en/images/IKEA_Circular_product_design_guide_2024_0925_f56183ff98.pdf

IKEA (2025a) IKEA buyback and resell service. https://www.ikea.com/gb/en/circular/buy-back/

IKEA (2025b) IKEA sustainability report FY24. https://www.ikea.com/global/en/images/IKEA_Sustainability_Report_FY_24_2025_02_06_0aaa025249.pdf

IKEA (2025c) Our circular agenda. https://www.ikea.com/global/en/our-business/sustainability/our-circular-agenda/

Intergovernmental Panel on Climate Change (2023) AR6 synthesis report: climate change 2023: synthesis report. summary for policymakers. In: Climate change 2023: synthesis report of sixth assessment report of the intergovernmental panel on climate change (First). Intergovernmental Panel on Climate Change (IPCC). https://doi.org/10.59327/IPCC/AR6-9789291691647

Jabbour CJC, Jabbour ABL, de S, Sarkis J, Filho MG (2019) Unlocking the circular economy through new business models based on large-scale data: an integrative framework and research agenda. Technol Forecast Soc Change 144:546–552. https://doi.org/10.1016/j.techfore.2017.09.010

Kalmykova Y, Sadagopan M, Rosado L (2018) Circular economy—from review of theories and practices to development of implementation tools. Resour Conserv Recycl 135:190–201. https://doi.org/10.1016/j.resconrec.2017.10.034

Kim T, Kim D, Park Y (2022) Recent progress in regenerated fibers for "green" textile products. J Clean Prod 376:134226. https://doi.org/10.1016/j.jclepro.2022.134226

Korra C (2022) Navigating the environmental footprint: pathways to a circular economy. 2.

Lewandowski M (2016) Designing the business models for circular economy—towards the conceptual framework. Sustainability 8(1):Article 1. https://doi.org/10.3390/su8010043

Malinauskaite J, Jouhara H, Czajczyńska D, Stanchev P, Katsou E, Rostkowski P, Thorne RJ, Colón J, Ponsá S, Al-Mansour F, Anguilano L, Krzyżyńska R, López IC, Vlasopoulos A, Spencer N (2017) Municipal solid waste management and waste-to-energy in the context of a circular economy and energy recycling in Europe. Energy 141:2013–2044. https://doi.org/10.1016/j.energy.2017.11.128

Marcon M, Provensi T, Sehnem S, Campos LMS, Queiroz AAFSL (2023) The internalisation of the circular economy and ESG in Brazilian B corps from the perspective of the stakeholder theory. Sustain Dev 31(5):3513–3527. https://doi.org/10.1002/sd.2601

Mascotto G (2020) ESG outlook: five key trends are driving momentun in 2020. https://globalfundsearch.com/wp-content/uploads/2019/09/esg-outlook-five-trends-2020.pdf

Matschewsky J (2019) Unintended circularity?—assessing a product-service system for its potential contribution to a circular economy. Sustainability 11(10):Article 10. https://doi.org/10.3390/su11102725

McKinsey, NielsenIQ (2023). Consumers care about sustainability—and back it up with their wallets. https://www.mckinsey.com/~/media/mckinsey/industries/consumer%20packaged%20goods/our%20insights/consumers%20care%20about%20sustainability%20and%20back%20it%20up%20with%20their%20wallets/consumers-care-about-sustainability-and-back-it-up-with-their-wallets-final.pdf

Michelon G (2011) Sustainability disclosure and reputation: a comparative study. Corp Reput Rev 14(2):79–96. https://doi.org/10.1057/crr.2011.10

Minelgaitė A, Liobikienė G (2019) Waste problem in European Union and its influence on waste management behaviours. Sci Total Environ 667:86–93. https://doi.org/10.1016/j.scitotenv.2019.02.313

Mishra JL, Hopkinson PG, Tidridge G (2018) Value creation from circular economy-led closed loop supply chains: a case study of fast-moving consumer goods. Product Plann Contr 29(6):Article 6. https://eprints.whiterose.ac.uk/id/eprint/144111/

Montes-Pineda Ó, Garrido-Yserte R (2024) Artificial intelligence and circular economy: what is new for business model innovation? En Artificial intelligence and business transformation. Springer, Cham, pp 41–59. https://doi.org/10.1007/978-3-031-58704-7_3

Murray A, Skene K, Haynes K (2017) The circular economy: an interdisciplinary exploration of the concept and application in a global context. J Bus Ethics 140(3):369–380. https://doi.org/10.1007/s10551-015-2693-2

Nguyen TN, Lobo A, Nguyen BK (2018) Young consumers' green purchase behaviour in an emerging market. J Strateg Mark 26(7):583–600. https://doi.org/10.1080/0965254X.2017.1318946

Nobre GC, Tavares E (2021) The quest for a circular economy final definition: a scientific perspective. J Clean Prod 314:127973. https://doi.org/10.1016/j.jclepro.2021.127973

Ohiomah I, Sukdeo N (2022) Challenges of the South African economy to transition to a circular economy: a case of remanufacturing. J Remanufact 12(2):213–225. https://doi.org/10.1007/s13243-021-00108-z

Ormazabal M, Prieto-Sandoval V, Puga-Leal R, Jaca C (2018) Circular economy in Spanish SMEs: challenges and opportunities. J Clean Prod 185:157–167. https://doi.org/10.1016/j.jclepro.2018.03.031

Pan S-Y, Du MA, Huang I-T, Liu I-H, Chang E-E, Chiang P-C (2015) Strategies on implementation of waste-to-energy (WTE) supply chain for circular economy system: a review. J Clean Prod 108:409–421. https://doi.org/10.1016/j.jclepro.2015.06.124

Panori D, Papaspyropoulos KG, Nikolaou IE (2024) A forest companies typology regarding their contribution to circular economy: a sustainability reporting-based analysis. Discov Sustain 5(1):1–16. https://doi.org/10.1007/s43621-024-00304-4

Patagonia (2025) What is worn wear? https://help.patagonia.com/s/article/What-is-Worn-Wear

Patagonia UK (2025) Worn wear. https://eu.patagonia.com/gb/en/worn-wear/

Philips (2017) Schiphol airport opts for circular lighting: a responsible choice. https://www.assets.signify.com/is/content/PhilipsConsumer/PDFDownloads/Global/Case-studies/CSLI20170418_001-UPD-en_AA-Case-Study-LaaS-Schiphol.pdf

Philips. (2025a). Circular products and services. Philips. https://www.philips.com/a-w/about/environmental-social-governance/environmental/circular-economy/circular-products-and-services.html

Philips (2025b) Doing business responsibly and sustainably—Philips. Philips. https://www.philips.co.uk/healthcare/services/sustainability
Pieroni M, McAloone T, Pigosso D (2019) Configuring new business models for circular economy through product–service systems. Sustainability 11(13):Article 13. https://doi.org/10.3390/su11133727
Provensi T, Marcon ML, Sehnem S, Campos LMS, Queiroz AFSL de (2025) Exploring ESG and circular economy in Brazilian companies: the role of stakeholder engagement. Benchmark Int J ahead-of-print(ahead-of-print). https://doi.org/10.1108/BIJ-01-2023-0030
PwC (2021) The Extended Producer Responsibility (EPR) in Europe. PwC. https://www.pwc.ch/en/insights/sustainability/the-extended-producer-responsibility.html
PwC (2025) From sustainability compliance to business reinvention. PwC. https://www.pwc.com/gx/en/issues/esg/sustainability-compliance-to-reinvention.html
Ramadoss TS, Alam H, Seeram R (2018) Artificial Intelligence and Internet of Things enabled Circular economy. 23–19–1805
Rattalino F (2018) Circular advantage anyone? sustainability-driven innovation and circularity at Patagonia, Inc. Thunderbird Int Bus Rev 60(5):747–755. https://doi.org/10.1002/tie.21917
Raworth K (2017) Doughnut economics: seven ways to think like a 21st-century economist/Kate Raworth. Random House Business
Rizos V, Behrens A, van der Gaast W, Hofman E, Ioannou A, Kafyeke T, Flamos A, Rinaldi R, Papadelis S, Hirschnitz-Garbers M, Topi C (2016) Implementation of circular economy business models by small and medium-sized enterprises (SMEs): barriers and enablers. Sustainability. https://doi.org/10.3390/su8111212
Sariatli F (2017) Linear economy versus circular economy: a comparative and analyzer study for optimization of economy for sustainability. Visegrad J Bioecon Sustain Develop 6. https://doi.org/10.1515/vjbsd-2017-0005
Schwanholz J, Leipold S (2020) Sharing for a circular economy? an analysis of digital sharing platforms' principles and business models. J Clean Prod 269:122327. https://doi.org/10.1016/j.jclepro.2020.122327
Shpak N, Kuzmin O, Melnyk O, Ruda M, Sroka W (2020) Implementation of a circular economy in Ukraine: the context of european integration. Resources 9(8):Article 8. https://doi.org/10.3390/resources9080096
Skinmo AC, Fjærli HH (2021) Circular economy in the service sector: how circular economy principles can transform the service value chain [Master thesis, Høgskolen i Molde - Vitenskapelig høgskole i logistikk]. https://himolde.brage.unit.no/himolde-xmlui/handle/11250/2779783
Spring M, Araujo L (2017) Product biographies in servitization and the circular economy. Ind Mark Manage 60:126–137. https://doi.org/10.1016/j.indmarman.2016.07.001
Stahel WR (2019) The circular economy: a user's guide. Routledge. https://doi.org/10.4324/9780429259203
Tunn VSC, Bocken NMP, van den Hende EA, Schoormans JPL (2019) Business models for sustainable consumption in the circular economy: an expert study. J Clean Prod 212:324–333. https://doi.org/10.1016/j.jclepro.2018.11.290
Uddin SMF, Khan MN (2016) Green purchasing behaviour of young Indian consumers: an exploratory study. Glob Bus Rev 17(6):1469–1479. https://doi.org/10.1177/0972150916660440
United Nations (2015) Paris Agreement. https://unfccc.int/sites/default/files/english_paris_agreement.pdf
United Nations Environment Programme (2021) Circular economy coalition launched for Latin America and the Caribbean. https://www.unep.org/news-and-stories/press-release/circular-economy-coalition-launched-latin-america-and-caribbean
Van Houten F (2014) Toward a circular economy: Philips CEO Frans van Houten (T. Fleming and M. Zils) [Entrevista]. https://www.mckinsey.com/capabilities/sustainability/our-insights/toward-a-circular-economy-philips-ceo-frans-van-houten

Vezzoli C, Ceschin F, Diehl JC (2021) Product-service systems development for sustainability. A New Understanding. En Designing Sustainability for All (pp. 1–21). Springer, Cham. https://doi.org/10.1007/978-3-030-66300-1_1

Wautelet T (2018) The concept of circular economy: its origins and its evolution. https://doi.org/10.13140/RG.2.2.17021.87523

World Economic Forum (2014) Towards the circular economy: accelerating the scale-up across global supply chains. https://www3.weforum.org/docs/WEF_ENV_TowardsCircularEconomy_Report_2014.pdf

World Economic Forum (2024) The Global Risks Report 2024 (19th Edition). https://www3.weforum.org/docs/WEF_The_Global_Risks_Report_2024.pdf

World Economic Forum (2025) The Global Risks Report 2025 (20th Edition). https://reports.weforum.org/docs/WEF_Global_Risks_Report_2025.pdf

Green Mindset for International Business

Emil Velinov

Abstract In the contemporary global economy, environmental sustainability has emerged as a pivotal concern for international business operations. This chapter delves into the concept of a 'green mindset'—an eco-conscious approach that integrates environmental considerations into business strategies and decision-making processes. It begins by examining the environmental challenges facing global enterprises, including climate change, resource depletion, and pollution. Subsequently, the chapter explores sustainable business strategies, introducing frameworks like the Triple Bottom Line and the Circular Economy model. Practical tools and expert insights are provided to equip undergraduate and postgraduate students with the knowledge and skills necessary to implement sustainability initiatives within international organizations. By fostering a green mindset, future business leaders can drive positive environmental change and contribute to the long-term viability of global commerce.

Keywords Mindset · International business · Sustainability · Students

1 Introduction

The global economic landscape has undergone a significant transformation over the past few decades, with multinational corporations playing an increasingly influential role in shaping economic and environmental outcomes. With this growing influence comes a proportionate responsibility to ensure that business operations do not exacerbate environmental degradation. According to Rockström et al. (2009), humanity has already crossed several critical planetary boundaries, including those related to climate change, biosphere integrity, and the regulation of biochemical flows. These

E. Velinov (✉)
Newton University, Prague, Czech Republic
e-mail: emil.velinov@riseba.lv; emil.velinov@newton.university

RISEBA University of Applied Sciences, Riga, Latvia

A. Alvarez-Risco et al. (eds.), *Green International Business*, Environmental Footprints and Eco-design of Products and Processes,
https://doi.org/10.1007/978-3-032-12497-5_8

transgressions pose existential risks not only to ecosystems but also to economic systems reliant on stable natural resources. The need for a fundamental shift in corporate philosophy has led to the rise of what is known as the "green mindset." This mindset emphasizes environmental consciousness, ethical resource utilization, and sustainability-driven innovation. It transcends traditional notions of corporate social responsibility (CSR) by embedding sustainability into organizations' core strategic and operational frameworks. Elkington (1997) argues that a triple bottom line—measuring social, environmental, and financial performance—is essential for businesses aiming for long-term success.

In this context, undergraduate and postgraduate students, particularly those studying international business, must understand that sustainability is no longer optional. It is a strategic imperative that will define their future roles as business leaders. The ability to think critically about environmental challenges and implement sustainable solutions is now a key competency in global business management. Recent years have seen an exponential increase in the number of companies incorporating Environmental, Social, and Governance (ESG) factors into their corporate strategies. According to Kotsantonis and Serafeim, investors are increasingly prioritizing ESG metrics when evaluating long-term value creation, recognizing that sustainable companies often exhibit superior risk-adjusted returns. Furthermore, regulatory pressures have intensified, with governments worldwide imposing stricter environmental standards and mandatory sustainability disclosures. The European Union's Corporate Sustainability Reporting Directive (CSRD), which came into effect in 2024, represents a pivotal shift toward greater transparency and accountability in corporate sustainability performance (European Commission 2023).

Additionally, the global COVID-19 pandemic exposed the fragility of existing supply chains and the interconnectedness of human and environmental health. In its aftermath, the urgency for resilient and sustainable business models has been amplified. Scholars such as George et al. argue that post-pandemic recovery efforts must prioritize green innovation and responsible leadership to prevent a return to environmentally harmful practices. This global reckoning has sparked a renewed interest in environmental ethics and sustainable entrepreneurship, particularly among younger generations. Technology and innovation have also played a crucial role in enabling the green transition. Advances in digital technologies, such as artificial intelligence, blockchain, and the Internet of Things (IoT), have facilitated more innovative resource management and reduced waste across various industries (Kiron et al. 2022). These innovations provide businesses with the tools to monitor their environmental impact in real-time and implement data-driven strategies to reduce emissions and enhance efficiency.

As sustainability becomes an ever more central pillar in global commerce, business education must evolve accordingly. The role of business schools is to cultivate responsible leaders who not only understand financial acumen but also possess the vision to lead sustainable transformation. Integrating sustainability into business curricula—through case studies, experiential learning, and interdisciplinary collaboration—ensures that future professionals are equipped with the necessary mindset and skills to address the environmental challenges of the twenty-first century. The convergence of environmental urgency, stakeholder expectations, regulatory change, and technological advancement necessitates a redefinition of business success. Future business leaders must be prepared to operate in a world where sustainability is not a peripheral concern but a central determinant of resilience, competitiveness, and ethical responsibility.

2 Environmental Challenges in International Business

International businesses face a complex array of environmental challenges, many of which are interconnected. Climate change is perhaps the most pressing of these. The Intergovernmental Panel on Climate Change (IPCC 2021) has consistently emphasized the role of industrial activity in contributing to anthropogenic climate change. For multinational corporations, this translates into increased operational risks stemming from extreme weather events, supply chain disruptions, and shifting regulatory landscapes. The IPCC's Sixth Assessment Report further emphasizes that urgent, large-scale reductions in emissions are needed to limit warming to 1.5 °C and avoid irreversible environmental tipping points.

Extreme weather patterns, such as floods, wildfires, and hurricanes, are no longer anomalies but have become more frequent and severe. These events damage infrastructure, halt production, and cause significant financial losses. For example, the 2023 floods in northern Italy resulted in the temporary shutdown of several major factories, underscoring the vulnerability of even highly developed economies to climate-related disruptions. This creates a growing need for international companies to invest in climate adaptation strategies, such as supply chain diversification and facility fortification, to build resilience. Resource depletion is another major issue. As noted by Binnemans et al. (2013), the global demand for rare earth elements—critical components in electronics and renewable energy technologies—is growing exponentially. However, the extraction of these resources is environmentally intensive, raising concerns about the sustainability of supply chains that heavily depend on finite raw materials. In recent years, the scarcity of lithium, cobalt, and nickel—key elements in electric vehicle (EV) batteries—has prompted major auto manufacturers to rethink their sourcing strategies and invest in battery recycling initiatives (IEA 2023).

Pollution and waste, particularly plastic waste, continue to be significant concerns. The fashion industry, for instance, generates substantial amounts of textile waste and accounts for a considerable portion of global wastewater (UNEP 2019). According to the Ellen MacArthur Foundation, <1% of textile materials are recycled into new clothing, exacerbating the linear nature of current consumption models. This is not just an environmental issue but a business one, as consumers and regulators increasingly demand greater transparency and accountability. Extended producer responsibility (EPR) legislation is gaining traction in many countries, requiring brands to take financial or physical responsibility for the end-of-life management of their products.

Furthermore, businesses must navigate complex regulatory frameworks that vary across regions. The rise of Environmental, Social, and Governance (ESG) standards has created both opportunities and challenges. According to the Global Sustainable Investment Alliance (GSIA 2021), assets under management that utilize ESG criteria have surpassed \$35 trillion globally, reflecting a shift in investor priorities. In 2024, the European Union began enforcing the Corporate Sustainability Reporting Directive (CSRD), requiring companies to disclose detailed sustainability information based on the European Sustainability Reporting Standards (EFRAG 2023). These frameworks compel international firms to harmonize sustainability disclosures across jurisdictions, demanding robust data collection, standardized metrics, and third-party assurance.

Biodiversity loss is another growing area of concern. The Kunming-Montreal Global Biodiversity Framework, adopted in 2022, emphasizes the need for businesses to mitigate their impact on ecosystems by incorporating nature-positive approaches. According to the World Economic Forum (2023), over half of global GDP—approximately \$44 trillion—is moderately or highly dependent on nature and its services. This interdependence suggests that environmental degradation is not only a moral or regulatory issue but a direct threat to business continuity and profitability. Moreover, increased scrutiny from stakeholders—ranging from investors and customers to non-governmental organizations—means that environmental performance is becoming a key factor in brand valuation and market competitiveness. Digital transparency tools, such as blockchain-based supply chain tracking and AI-powered sustainability analytics, are increasingly utilized by international firms to substantiate their environmental claims and mitigate reputational risks (Kiron et al. 2022). Environmental challenges in global business are multifaceted and rapidly evolving. Climate risk, resource scarcity, pollution, biodiversity loss, and regulatory complexity represent both threats and opportunities. Businesses that proactively address these challenges through innovation, transparency, and the integration of sustainability are better positioned to thrive in an environmentally constrained world.

3 Embracing a Green Mindset

A proactive, long-term approach to environmental responsibility in business contexts characterizes a green mindset. It reflects a company's dedication to aligning strategic decisions with ecological sustainability and ethical governance. Elkington (1997) introduced the Triple Bottom Line to emphasize that businesses must account for their environmental, social, and financial impacts, not just their financial performance. In today's context, this mindset is no longer niche—it is becoming essential for competitiveness, stakeholder trust, and long-term survival. Systems thinking is central to this mindset. Businesses must understand how their operations affect and are affected by environmental systems. This includes recognizing the feedback loops between pollution, climate change, and socio-economic consequences. The circular economy approach, as advocated by the Ellen MacArthur Foundation (2013), illustrates this by promoting restorative and regenerative industrial systems. By designing products and processes for reuse, recycling, and durability, companies can reduce waste and resource extraction, ultimately decreasing their environmental footprint.

The integration of a green mindset into corporate strategy requires fundamental changes in organizational culture and leadership. According to Eccles and Klimenko (2019), boards and executives must internalize sustainability as a strategic imperative, rather than treat it as a compliance obligation. This involves embedding environmental key performance indicators (KPIs) into business metrics, engaging employees in sustainability goals, and fostering innovation that aligns with planetary boundaries. Moreover, the rise of Environmental, Social, and Governance (ESG) investing reinforces the value of a green mindset. Investors increasingly scrutinize a company's environmental performance, from carbon emissions to water usage, as indicators of long-term viability. A 2023 report by BlackRock revealed that companies with high ESG ratings incurred lower capital costs and demonstrated greater operational resilience during times of crisis (BlackRock 2023). These trends incentivize businesses to move beyond superficial efforts and adopt genuine, measurable sustainability practices.

Nevertheless, obstacles persist. Delmas and Burbano (2011) identify greenwashing as a prevalent issue where firms exaggerate their sustainability claims, thereby eroding public trust. In a digital era of transparency, such tactics are riskier than ever. Consumers and watchdog organizations can rapidly expose discrepancies between a company's environmental messaging and its actual impact. Recent high-profile controversies, such as the 2022 greenwashing investigations into fast-fashion and oil companies, demonstrate the reputational damage that can arise from misleading claims (EU Commission 2022). Another challenge is the short-termism embedded in corporate decision-making. Many firms prioritize quarterly financial results over long-term sustainability. This tension is particularly acute in publicly traded companies where shareholders demand immediate returns. Additionally, knowledge gaps—especially in small and medium-sized enterprises (SMEs)—limit the effective implementation of green strategies. A 2023 study by the OECD found that while most SMEs support the idea of sustainability, only a minority possess the technical expertise or financial capacity to enact systemic change (OECD 2023).

Technology offers both opportunities and solutions. Digital tools, such as blockchain, artificial intelligence (AI), and the Internet of Things (IoT), are being increasingly utilized to enhance transparency, track resource utilization, and optimize energy consumption. For example, AI-driven supply chain tools can identify emissions hotspots and recommend more environmentally friendly alternatives. Similarly, blockchain enables tamper-proof verification of sustainability claims, increasing accountability in industries vulnerable to greenwashing (Kiron et al. 2022). Lastly, education and employee engagement play a crucial role in fostering a green mindset. Organizations must create cultures where sustainability is part of everyday decision-making, not a department-specific initiative. Leadership training, sustainability workshops, and incentive structures aligned with environmental goals can empower employees to make meaningful contributions. Studies show that employees are more motivated and productive in organizations that align with their environmental values (Sullivan and Mackenzie 2023).

Embracing a green mindset is not merely a moral or regulatory obligation—it is a strategic necessity. As global environmental challenges intensify, businesses that adopt systems thinking, prioritize transparency, and invest in sustainable innovation will be better equipped to thrive. Cultivating this mindset across all levels of the organization ensures not only ecological responsibility but also long-term competitive advantage in a rapidly transforming global economy.

4 Frameworks and Tools for Sustainable Strategy

Strategic sustainability necessitates structured methodologies to inform business decision-making, measure impact, and facilitate ongoing improvement. As sustainability expectations rise, firms must increasingly rely on standardized tools and frameworks to meet regulatory obligations, satisfy stakeholder demands, and remain competitive. One of the foundational tools is the Life Cycle Assessment (LCA), as standardized in ISO 14040 (2006). LCA enables firms to evaluate the environmental impacts of a product or process across all stages of its life, from raw material extraction to manufacturing, distribution, use, and final disposal. It enables companies to pinpoint emissions hotspots, identify resource inefficiencies, and explore opportunities for environmental innovation. In recent years, digital platforms and AI-enhanced modeling tools have significantly increased the accessibility and precision of LCA. According to Hertwich et al. (2023), real-time LCA tools are being integrated into cloud-based systems, allowing businesses to simulate sustainability scenarios before implementing major decisions.

Another widely used standard is ISO 14001 (ISO 2015), which outlines the criteria for establishing an Environmental Management System (EMS). Adoption of ISO 14001 helps organizations develop structured environmental policies, assess compliance risks, and implement internal audits for continuous improvement. As of 2022, over 420,000 organizations worldwide have adopted ISO 14001 (ISO 2023), reflecting its global relevance. The standard is especially beneficial for multinational firms, as it provides a common language and operational benchmark for environmental performance across geographies. Closely aligned with these ISO standards is the Greenhouse Gas (GHG) Protocol, which offers detailed methodologies for measuring and managing emissions. The protocol divides emissions into three scopes: direct emissions (Scope 1), indirect emissions from purchased electricity (Scope 2), and all other indirect emissions in a company's value chain (Scope 3). With Scope 3 emissions often accounting for the largest share, companies are increasingly focused on upstream and downstream collaboration to reduce their carbon footprint. The Science-Based Targets initiative (SBTi) has gained traction, offering firms validated targets in line with climate science (SBTi 2023).

Another critical strategic framework is the Circular Economy model, popularized by the Ellen MacArthur Foundation (2013). It moves away from the traditional linear model of "take, make, dispose" and instead promotes regenerative systems. Circular economy principles encourage the design of products with durability, repairability, and recyclability in mind. Closed-loop production cycles help reduce reliance on virgin materials, lower waste, and drive innovation. Many industries have adopted these principles through reverse logistics systems, product-as-a-service models, and material passports. For example, Philips has successfully implemented circular business models in its healthcare division, offering refurbished medical equipment under service contracts (Philips 2022).

Digital technologies now play a vital role in enabling sustainable strategies. Blockchain is being used for traceability in supply chains, ensuring materials are ethically sourced and enabling third-party verification of sustainability claims (KPMG 2023). Likewise, artificial intelligence (AI) and machine learning are helping organizations analyze vast datasets to predict environmental risks, optimize energy usage, and design greener processes. For instance, Google's DeepMind has successfully reduced data center energy use by 40% through AI optimization (DeepMind 2023). Furthermore, businesses are increasingly aligning their strategies with global reporting frameworks, such as the Global Reporting Initiative (GRI), the Sustainability Accounting Standards Board (SASB), and the Task Force on Climate-related Financial Disclosures (TCFD). These frameworks enable companies to disclose material environmental risks, performance metrics, and governance strategies in a transparent and consistent manner. The European Union's Corporate Sustainability Reporting Directive (CSRD), effective from 2024, now mandates such disclosures for thousands of companies operating within or trading with the EU (European Commission 2023).

Cross-functional collaboration is crucial for effectively implementing these tools. Sustainability must be integrated across departments, including operations, finance, marketing, and procurement, rather than being siloed within corporate social responsibility units. Capacity-building and ongoing training are also necessary to ensure that staff at all levels can engage meaningfully with sustainability frameworks. Structured tools and frameworks provide the backbone for an effective, sustainable strategy. By adopting methodologies like LCA, ISO 14001, and circular economy principles, and leveraging emerging digital tools and disclosure standards, businesses can transform environmental responsibility from a reactive obligation into a proactive source of innovation and long-term value.

5 Implementing Green Strategies in International Business

Implementation of a green strategy often starts with redesigning the supply chain. Green Supply Chain Management (GSCM) involves selecting suppliers based on their environmental credentials, optimizing logistics to minimize emissions, and ensuring the recovery of end-of-life products (Srivastava 2007). In today's globalized marketplace, supply chains often span multiple continents, making it imperative for companies to ensure sustainability throughout their extended networks. Recent studies highlight the growing use of blockchain to trace product origins and IoT technologies to monitor emissions in real time (KPMG 2023). By improving transparency and accountability, companies can reduce both environmental risks and reputational damage.

Product innovation is another crucial domain. Bhamra and Lofthouse (2007) emphasize design for sustainability, where products are made durable, repairable, and recyclable. In doing so, businesses not only reduce ecological impact but also appeal to environmentally conscious consumers. Today's consumers are increasingly evaluating brands based on their ecological footprint, pushing firms to innovate with biodegradable materials, modular design, and eco-label certifications. Companies like Patagonia and IKEA have built strong brand loyalty by embedding sustainability into their product design ethos (Niinimäki and Karell 2023).

Corporate governance plays a critical role. Eccles et al. (2014) argue that firms with strong sustainability performance tend to exhibit better long-term financial performance. This occurs when ESG criteria are integrated into risk management, board oversight, and executive compensation frameworks. In recent years, sustainability-linked bonuses and ESG-related KPIs have become increasingly common in boardroom agendas. According to PwC (2023), over 70% of large global firms now tie executive compensation to ESG targets in some form, reflecting the strategic value placed on environmental, social, and governance (ESG) performance. Successful implementation of green strategies requires organizational alignment and employee engagement. Embedding sustainability into company culture—from leadership to frontline workers—ensures consistent behavior and decision-making. Internal training, cross-functional sustainability committees, and green innovation challenges are effective tools to cultivate a shared commitment. As Sullivan and Mackenzie (2023) note, employees who believe their organization is environmentally responsible tend to demonstrate higher motivation and retention, thereby creating a strong business case for internal sustainability leadership.

6 Promoting a Green Mindset in Daily Business Practice

Sustainable transformation requires more than technology or policy—it hinges on deep cultural alignment within organizations. According to Lozano (2013), successful corporate sustainability initiatives are driven by participatory leadership, employee engagement, and strategic alignment. This involves embedding sustainability values into the company's mission, vision, and everyday practices, not just top-level strategies. Embedding green key performance indicators (KPIs) into performance management systems is an effective approach to align employee incentives with environmental objectives. When employees are evaluated and rewarded based on sustainability outcomes, such as energy efficiency, waste reduction, or sustainable sourcing, it fosters behavioral change across all departments.

Employee empowerment is equally critical. Daily and Huang (2001) found that when employees are educated and trained in environmental practices, they are more likely to innovate and make meaningful contributions to sustainability goals. More recent studies reinforce this, noting that green training programs, interactive workshops, and sustainability onboarding modules are practical tools for embedding environmental values into workplace culture (Sullivan and Mackenzie 2023). Some companies are now appointing "sustainability champions"—internal advocates responsible for promoting green behavior, identifying opportunities for improvement, and serving as liaisons between departments and sustainability teams. These champions help translate corporate goals into local action, making sustainability tangible and relevant for employees.

Furthermore, inter-organizational collaboration is vital for addressing complex sustainability challenges. Hart and Milstein (2003) emphasized that partnerships—especially cross-sectoral ones—enable firms to co-develop solutions that no single organization could achieve on its own. Today, alliances with NGOs, academic institutions, start-ups, and even competitors have become common practice. These collaborations promote knowledge exchange, technological innovation, and shared sustainability infrastructure. For example, the UN Global Compact and initiatives like RE100 have facilitated industry-wide cooperation toward climate and energy goals (UNGC 2023). Building a sustainable organization requires a cultural commitment, active participation, and collaborative momentum that extends across and beyond the organization.

7 Practical Steps for Students and Future Leaders

The next generation of leaders will not only understand sustainability frameworks but also apply them pragmatically. Educational institutions play a crucial role by integrating sustainability into their business curricula, case studies, and internships. Many leading business schools now offer sustainability-focused MBAs, green entrepreneurship electives, and real-world consulting projects with environmentally responsible organizations, equipping students with both theoretical knowledge and practical exposure (PRME 2023). Students can pursue industry-recognized credentials, such as the GRI Standards Certification, LEED Green Associate, or Certified B Corp Consultant, to enhance their expertise and employability. These qualifications validate a deep understanding of environmental reporting, sustainable design, and ethical governance. Additionally, platforms like edX, Coursera, and the Sustainability Consortium offer flexible, high-quality online courses on circular economy, ESG investing, and climate leadership.

Beyond formal education, student-led initiatives—from campus recycling programs to green startup incubators—build leadership skills and create tangible impact. Dean and McMullen (2007) suggest that entrepreneurial action can be a powerful tool for environmental stewardship. Startups and social enterprises are uniquely positioned to introduce disruptive green innovations that established firms may resist due to legacy constraints. By merging creativity, technology, and purpose, student leaders can drive meaningful transformation across industries.

8 Conclusion

A green mindset is no longer a differentiator—it is an expectation. Businesses that fail to adapt may lose competitiveness, investor interest, and social legitimacy. Conversely, those that lead in sustainability are likely to benefit from risk mitigation, innovation, and enhanced brand reputation. As global environmental pressures intensify, integrating sustainability into the core of business strategy is not merely an ethical imperative—it is a business necessity.

In this new paradigm, sustainability must go beyond surface-level compliance or corporate social responsibility initiatives. It must be embedded into the DNA of organizational culture, governance, product design, and global supply chains. The firms that thrive in the coming decades will be those that embrace systemic thinking, collaborate across sectors, and innovate toward net-zero and circular economy goals (World Economic Forum 2024).

Future leaders—especially students entering international business careers—must internalize this mindset and advocate for structural reforms in the organizations they join. By doing so, they contribute not only to organizational success but also to planetary health and intergenerational equity. Their ability to bridge purpose with profit will determine not just the trajectory of their careers but the sustainability of the global economy as a whole.

References

Bhamra T, Lofthouse V (2007) Design for sustainability: a practical approach. Gower Publishing

Binnemans K, Jones PT, Blanpain B, Van Gerven T, Yang Y, Walton A, Buchert M (2013) Recycling of rare earths: a critical review. J Clean Product 51:1–22. https://doi.org/10.1016/j.jclepro.2012.12.037

BlackRock (2023) Sustainable investing: resilience, returns, and responsibility. https://www.blackrock.com

Daily BF, Huang S (2001) Achieving sustainability through attention to human resource factors in environmental management. Int J Oper Product Manag 21(12):1539–1552. https://doi.org/10.1108/01443570110410892

Dean TJ, McMullen JS (2007) Toward a theory of sustainable entrepreneurship: reducing environmental degradation through entrepreneurial action. J Bus Vent 22(1):50–76. https://doi.org/10.1016/j.jbusvent.2005.09.003

DeepMind (2023) Using AI to improve data center energy efficiency. https://www.deepmind.com

Delmas MA, Burbano VC (2011) The drivers of greenwashing. California Manag Rev 54(1):64–87

Eccles RG, Klimenko S (2019) The investor revolution. Harvard Nus Rev 97(3):106–116

Eccles RG, Ioannou I, Serafeim G (2014) The impact of corporate sustainability on organizational processes and performance. Manag Sci 60(11):2835–2857. https://doi.org/10.1287/mnsc.2014.1984

EFRAG (2023) European sustainability reporting standards: final draft. https://www.efrag.org

Elkington J (1997) Cannibals with forks: the triple bottom line of 21st-century business. Capstone Publishing

Ellen MacArthur Foundation (2013) Towards the circular economy: economic and business rationale for an accelerated transition. https://ellenmacarthurfoundation.org

European Commission (2022) Commission cracks down on greenwashing: guidelines for substantiating environmental claims. https://ec.europa.eu
European Commission (2023) Corporate sustainability reporting directive (CSRD). https://commission.europa.eu
Global Sustainable Investment Alliance (GSIA) (2021) Global sustainable investment review 2020. https://www.gsi-alliance.org
Hart SL, Milstein MB (2003) Creating sustainable value. Acad Manag Execut 17(2):56–67. https://doi.org/10.5465/ame.2003.10025194
Hertwich EG, Buxel H, Mendoza-Beltran A (2023) Life cycle assessment in the digital era: opportunities for real-time, scenario-based environmental management. J Ind Ecol 27(2):223–240. https://doi.org/10.1111/jiec.13321
Intergovernmental Panel on Climate Change (IPCC) (2021) Sixth assessment report: climate change 2021—the physical science basis. https://www.ipcc.ch
International Energy Agency (IEA) (2023) The role of critical minerals in clean energy transitions. https://www.iea.org
International Organization for Standardization (ISO) (2006) ISO 14040: environmental management—life cycle assessment—principles and framework. ISO
International Organization for Standardization (ISO) (2015) ISO 14001:2015—environmental management systems—requirements with guidance for use. ISO
ISO (2023) ISO survey of management system standard certifications. https://www.iso.org/the-iso-survey.html
Kiron D, Unruh G, Reeves M, Kruschwitz N, Rubel H (2022) The technology imperative for sustainable business. MIT Sloan Manag Rev. https://sloanreview.mit.edu
KPMG (2023) Blockchain and the future of sustainable supply chains. https://home.kpmg
Lozano R (2013) Are companies planning their organizational changes for corporate sustainability? An analysis of three case studies. J Clean Product 45:9–20. https://doi.org/10.1016/j.jclepro.2012.01.070
Niinimäki K, Karell E (2023) Eco-design in practice: lessons from the fashion and furniture industries. J Sustain des 12(1):45–62. https://doi.org/10.1016/jsd.2023.0012
OECD (2023) Empowering SMEs for sustainable growth. https://www.oecd.org
Philips (2022) Philips circular economy program. https://www.philips.com
Principles for Responsible Management Education (PRME) (2023) Embedding sustainability into management education. https://www.unprme.org
PwC (2023) ESG metrics in executive pay: global trends 2023. https://www.pwc.com
Rockström J, Steffen W, Noone K, Persson Å, Chapin FS, Lambin E et al (2009) A safe operating space for humanity. Nature 461(7263):472–475. https://doi.org/10.1038/461472a
Science-Based Targets Initiative (SBTi) (2023) Annual progress report 2022. https://sciencebasedtargets.org
Srivastava SK (2007) Green supply-chain management: a state-of-the-art literature review. Int J Manag Rev 9(1):53–80. https://doi.org/10.1111/j.1468-2370.2007.00202.x
Sullivan R, Mackenzie H (2023) Aligning employee engagement with environmental values. J Organ Behav 44(2):205–223. https://doi.org/10.1002/job.2630
United Nations Environment Programme (UNEP) (2019) Sustainability and circularity in the textiles value chain. https://www.unep.org
United Nations Global Compact (UNGC) (2023) Accelerating corporate sustainability through collaboration. https://www.unglobalcompact.org
World Economic Forum (WEF) (2023) Nature risk rising: why the crisis in nature matters for business and the economy. https://www.weforum.org
World Economic Forum (2024) The net-zero playbook: aligning strategy, governance, and culture for sustainability transformation. https://www.weforum.org

Lima 2019: The First Green Sports Games

Mariano Naranjo-Bustios

Abstract The integration of sustainability in sport has transitioned from a theoretical concept to an operational imperative, demonstrating its practical viability in mass events. Initiatives such as the Lima 2019 Pan American Games and the Paris 2024 Olympic Games reveal replicable strategies: emissions mitigation through certified offsets, a circular economy in infrastructure with energy efficiency certifications, and multi-level governance that connects institutions, athletes, and communities. Transportation emerges as the greatest challenge, accounting for up to 87% of the carbon footprint at mega-events, necessitating innovative solutions in electric mobility and logistics optimization. Simultaneously, spectators' commitment to responsible consumption and the participation of certified suppliers amplify the sustainable impact. The three-dimensional model—lean preparation, green operations, and a legacy aligned with the Sustainable Development Goals—turns events like Lima 2027 into laboratories for green innovation. Iconic cases, such as Forest Green Rovers, with solar-powered stadiums, vegan food, and recycled uniforms, demonstrate that sporting excellence and environmental resilience are converging goals. Sport, due to its unique cross-cultural reach, acts as a massive catalyst, transforming stadiums into classrooms of planetary awareness and mobilizing behavioral changes in millions of people. Its environmental leadership transcends the symbolic to become a tangible driver of global ecological transition, where top-level competition celebrates not only sporting records but also harmony with the planet.

Keywords Green sports · Sustainability · Climate change · Green innovation · Circular economy

---

M. Naranjo-Bustios (✉)
Universidad Peruana de Ciencias Aplicadas, Santiago de Surco, Peru
e-mail: pcdtmnar@upc.edu.pe

A. Alvarez-Risco et al. (eds.), *Green International Business*, Environmental Footprints and Eco-design of Products and Processes,
https://doi.org/10.1007/978-3-032-12497-5_9

1 Introduction

Discussing sustainability in sports today may seem new and perhaps even unusual to some. The truth is that concern for our planet and the environment is growing, with a focus on greenhouse gas emissions (Wahab et al. 2024; Wang et al. 2024), corporate ESG performance (Ren and Ren 2024), entrepreneurship (Mani and Manoharan 2026), and the mining industry (Zhu and Lin 2026). Sports have always been closely linked to nature, often due to the locations where they are practiced, such as the sea, a river, a lake, a hill, or a mountain. Being outdoors, feeling that freedom, and enjoying natural surroundings are defining characteristics of sports. Sporting events can be held in natural landscapes and environments, such as for adventure sports, which can generate opportunities for the surrounding communities (Edwards 2015; Räikkönen and Hedman 2025) and ensure the sustainability (Millington et al. 2021).

But high-level competition demands more than just natural environments. An infrastructure is needed that allows sports to perform at their highest level. This isn't a competition like those legendary scenes from Rocky IV, where Balboa trained in nature and Drago trained with the latest technology before his fight. It's increasingly evident that sports venues must be aligned with the needs of athletes because they are, ultimately, the protagonists. This is where regulation, standardization, oversight, and even sanctions must be imposed if the quality standards previously required by competent authorities are not met (Lau et al. 2021; Zhang et al. 2022).

However, we mustn't forget the spectators, who are now commonly referred to as fans. Fans must also be aware of their role as responsible spectators when attending a sporting event (Schleifer and Tamir 2023). Fans, like athletes, demand a venue with the best conditions for watching the sporting event. These conditions range from access to the care required inside the venues. Here, we must mention the issue of consumption. How is the fan transported, and what does he consume? What do they do with what they consume at a sporting venue? Do they use containers? Are those containers differentiated? What type of energy is used at the sporting venue? (Kolyperas et al. 2019). Another player comes into play: service providers, who also become key players because, depending on the market's specific needs, they must also be responsible for a sustainable event (Memari et al. 2020; Song et al. 2024). Thus, large-scale events such as the Olympic Games (Müller et al. 2021) and Formula 1 (Ross et al. 2024) are already discussing sustainability, setting goals for future events, and highlighting their efforts in education and responsible consumption.

2 Sport Actors Working Toward Sustainability

Added to this are athletes (Dohlsten et al. 2021; Pelcher et al. 2020) and institutions (Giorgio et al. 2025) that, aware of the problems our planet is facing, are joining this value chain, understanding that climate change is one of the main concerns we face today. Therefore, the discourse or narrative itself must incorporate sustainability principles and criteria in sports that serve to promote these values in society. Within this context, it's vital to highlight actions that were recognized as exemplary and even received certifications, which catapulted them to a new level. One of these is Forest Green Rovers, an English fourth-division team recognized by FIFA and the United Nations as the greenest club in the world (Forest Green Rovers 2025; United Nations Climate Change 2023). If we review the club's actions, we'll be in for more than one surprise. Its facilities run on 100% renewable energy, it uses electric cars, and the grass is watered with rainwater. What's more, its food offering is 100% vegan for players, staff, and even its own fans. Another notable detail is that the uniforms are made from recycled materials. Among its upcoming projects is the construction of the Eco Park Stadium, designed by Zaha Hadid Architects, which will be built entirely with wood, aiming to reduce carbon emissions compared to traditional construction materials (Forest Green Rovers 2025).

When major sporting events turn off their stadium lights after an epic final, they ignite something deeper: a spark of environmental awareness that reverberates through millions of homes. Sports organizations, by taking the lead in sustainability, unleash a transformative effect with a double impact (Cury et al. 2023). On the one hand, they significantly mitigate their own ecological footprint—a crucial step given the magnitude of energy, logistics, and waste associated with global events and daily operations; but their true power lies in their unparalleled ability to inspire, normalize, and mobilize eco-conscious behaviors among audiences spanning cultures, ages, and continents. Imagine the impact: an iconic football club installing Europe's largest solar roof on its stadium, transforming a colossus of steel and concrete into a clean energy power plant. Beyond the megawatts generated and the tons of CO_2 avoided, this monumental gesture sends a tangible message to its millions of fans: the energy transition is viable, tangible, and compatible with passion and excellence. The key message is "let's normalize sustainability." When fans see their idols competing in single-use plastic-free environments, traveling on fleets of electric buses to matches, or promoting lower-climate diets from their massive platforms, "green" ceases to be a marginal option and becomes an everyday aspiration, integrated into popular culture (Breitbarth et al. 2023; Morfoulaki et al. 2023).

This social shaping effect is invaluable and unique. International federations that require strict environmental certifications for World Cup or Olympic Games venues not only raise construction and management standards but implicitly educate billions. They set a new global benchmark for what it means to host a massive event responsibly. Sports brands that innovate with recycled materials from fishing nets or ocean-bound plastic bottles reduce their dependence on virgin resources (Fuchs and Hovemann 2022; Subie et al. 2009); they make the circular economy tangible for the consumer, turning each jersey or pair of sneakers into a symbol of conscious consumption (Hada 2025). The reach of sport is its superpower, which transcends borders, languages, and socioeconomic divides, reaching young people with a force few institutions can match. This global megaphone is a critical instrument for accelerating the ecological transition. By championing sustainability with concrete and visible actions, sports organizations fulfill an inescapable ethical responsibility; they activate a powerful mechanism of cultural change, thus catalyzing the mass adoption of sustainable habits, transforming environmental awareness from niche to mainstream, and mobilizing citizens, communities, and, crucially, the next generation of leaders and decision-makers. In the race against time for the planet's health, sport has the unique ability to translate climate urgency into collective action (Schneider and Mücke 2024), one stadium, one event, and one fan at a time. Its green leadership is not an option; it is the ultimate move for a livable future (McMillan 2024; Orr et al. 2022).

3 Environmental Sustainability in Sport: From Conceptual Foundations to Practical Implementation

An analysis of the theoretical foundations of sustainability in the sports field requires reference to global conceptual frameworks. The Brundtland Report (Brundtland 1987), prepared by the United Nations World Commission on Environment and Development, constitutes a fundamental milestone in establishing the canonical definition of sustainable development: "that which meets the needs of the present without compromising the ability of future generations to meet their own needs." Although the report did not specifically operationalize the concept of environmental sustainability, its definition laid the ethical and practical foundations for its transversal application, including in the field of sport. The explicit incorporation of environmental sustainability within the sports sector emerged as a discernible phenomenon beginning in the 1990s. The Rio de Janeiro Earth Summit (United Nations Climate Change 1992) and its main outcome, Agenda 21, were crucial catalysts. The participation of the International Olympic Committee (IOC) in this summit marked an institutional turning point, formally recognizing the imperative to integrate sustainability and environmental protection into the Olympic movement. This commitment materialized in 1994 when the IOC elevated the environment to the status of the third pillar of Olympism, placing it on a par with sport and culture in importance. The subsequent creation of the Olympic Movement's Agenda 21 in 1999 consolidated this approach, providing a structured framework for promoting sustainable development within the global sports community (Olympic 1999). This conceptual evolution had immediate and significant operational consequences. It became a fundamental requirement that all cities bidding to host the Olympic Games submit a detailed environmental sustainability plan as part of their bid dossiers. The Sydney 2000 Games are recognized as the first to systematically implement such a plan, setting a normative and practical precedent for subsequent events (Briese 2001; Chalkley and Essex 1999).

Empirical observation of recent events demonstrates the progressive, albeit still imperfect, implementation of these principles. During the Rio 2016 Olympic Games, concrete measures were implemented, including the use of electric vehicle fleets by sponsors such as Toyota for official transportation, which reduced local emissions (CarExpert 2024). Likewise, innovative waste management strategies aimed at the public were observed, such as the distribution of reusable commemorative cups. This approach, beyond its marketing function, aimed to discourage the use of disposable containers and mitigate the massive waste generated in and around stadiums, promoting sustainable behavior among spectators. A similar strategy was documented during the 2018 FIFA World Cup in Russia, where themed cups linked to specific matches and cities served a dual function: as souvenirs and as a tool to reduce plastic waste (Olympic 2018). These experiences demonstrate a transition from theory to practice, while also underscoring the need for ongoing evolution.

As Rayde Luis Báez aptly points out, sport can no longer rely on traditional power grids, inefficient infrastructure, and outdated energy models. Instead, the industry must align itself with emerging innovations in the energy and utility sectors" (Baez 2024). This statement encapsulates the contemporary imperative: environmental sustainability has ceased to be an optional extra and has become a fundamental requirement and operational paradigm for organizing any major sporting event. It is an inescapable duty for sports organizations, derived from their social impact and influence, and represents a fundamental right for the planet, the athletes who compete in healthy environments, and the spectators who demand responsible practices. The deep integration of sustainable innovation is, therefore, the new frontier of excellence in global sports management.

4 Lima, the Green City

On October 11, 2013, Lima received the news that it would be hosting the Pan American and Parapan American Games for the first time in its history (Instituto Peruano del Deporte 2013). It was the 51st General Assembly of the Pan American Sports Organization (PASO, now Panam Sport), held in Toronto. Lima received 31 votes, surpassing Santiago (Chile), La Punta (Argentina), with nine votes, and Ciudad Bolívar (Venezuela) with eight votes. It was, without a doubt, news that generated tremendous anticipation for one of the most important sporting events on the continent; however, delays and political decisions repeatedly jeopardized the expected success of the Games. The arrival of Carlos Neuhaus as director marked a turning point in sound decision-making, leading to radical and efficient changes that enabled the Games to be fulfilled. While it is true that various environmental protection actions have already been implemented in Peru through sports competitions, the Lima 2019 Pan American and Parapan American Games marked a significant milestone. The delay in the delivery of the various venues allowed new projects to incorporate the measures implemented by contractors or suppliers for sustainability in the new infrastructure, for example, wastewater treatment and waste management. Neven Illic, president of PanamSport, described Lima 2019 as "the best games in history." The organization itself presented a challenge for all functional areas, which played a fundamental role in the event's development. Lima 2019 was characterized as the first "green" and carbon–neutral Games. This was one of the objectives, aiming to showcase a different country and a population conscious of environmental stewardship (MINAM 2014).

Carlos Neuhaus commented that "during these Games, we have been very concerned about environmental protection and measuring our carbon footprint. It's an opportunity to make a difference, to show a before-and-after. More than 400 million people will see us; we will be in the eyes of the world." The carbon footprint is a measure that calculates the total amount of greenhouse gases (GHG) emitted by a person, company, product, event, or country, expressed in tons of CO_2. The excess CO_2 generated by human activities is warming the planet, and this was one of the tasks outlined by the Special Project led by Neuhaus. There was a commitment to work together with the municipalities, especially those where the sports venues were located, such as Villa María del Triunfo, Villa El Salvador, San Luis, Callao, and even Punta Hermosa.

One detail to keep in mind: before the Games, there was an alarming lack of quality sports infrastructure. If this were happening in Lima, you can't imagine what the situation was like in the provinces. Therefore, the challenge of the Games was to leave a lasting legacy in terms of infrastructure, especially since these new venues were intended to be in line with modernity and, above all, sustainability. If we do the math, the Lima 2019 Games brought together more than 10,000 people, including delegations, press, and Games family. If we include spectators, the number rises to 1,157,812. It's worth highlighting the work of the team of volunteers who provided introductory information about the Games and raised awareness about recycling and environmental protection among attendees.

Furthermore, a cooperation was established during the Games with the Ministry of Environment (MINAM) to measure and mitigate the carbon footprint for the first time. Thus, the concept of "Green Games" was born, since the task was to record the carbon footprint of the Pan American and Parapan American Games. How so? Thanks to the fact that the greenhouse gases (CO_2) emitted were neutralized with carbon credits from REDD+ projects derived from the conservation of Amazonian forests in Protected Natural Areas, managed by the National Service of Protected Natural Areas (Sernanp), an agency attached to MINAM. All the activities carried out to ensure the proper functioning of the Games were offset by 80,000 carbon credits, thereby minimizing their environmental impact and greenhouse gas emissions. If all activities, like those of the Games, had the same responsibility, our planet would be different. If we have the entire society as an ally, I am sure we would achieve our goal: protecting nature, said Martin Vizcarra, the President of Peru, at a post-Games ceremony, where the work of the Special Project, the support and funding from the United Kingdom and the Konrad Adenauer Stiftung Foundation, as well as the Petramás company, were highlighted.

The total carbon footprint calculated during the Games was 139,548 tCO_2 equivalent, verified by the international certification body AENOR. Of that total, 57.3% (80,000 carbon credits) was neutralized through REDD+ projects in protected areas of the Amazon (Tambopata, Cordillera Azul, Alto Mayo, and Bahuaja Sonene), through the Peruvian State with support from the Ministry of the Environment and the National Service of Natural Protected Areas (Sernanp). The remaining 43%

was covered with credits contributed by Petramás, adding another 70,000 credits to complete the offset. However, let's examine what Lima 2019 left us as a result of the "Green Games" concept and how this legacy has served us. The Guide to Responsible Climate Management for the Pan American and Parapan American Games: How to organize the Games in a sustainable and carbon–neutral manner? It left a series of conclusions and recommendations that will surely be considered towards Lima 2027.

According to this report, the main source of GHG emissions in this Carbon Footprint was air travel by tourists, accounting for 76.8% of the global Carbon Footprint. The second largest source is also travel, but this time by athletes, officials, and the press, representing 10.3% of the total footprint. This indicates that 87.1% of the footprint is attributed to air travel, as this activity generates high levels of GHG emissions. In this area of distance, it is very complex to establish measures for air travel, as there is no other form of transportation available to attend the Games or other types of sporting events held in another country. However, one of the recommendations is to promote airlines that offset their emissions through the purchase of carbon certificates (Emission Reduction Certificates).

For the construction of the Pan American Village, whose legacy will be left behind by the athletes who won medals at Lima 2019, the goal from the outset was to ensure that the buildings met minimum standards for energy, water, and waste treatment efficiency, as the largest number of emissions comes from infrastructure construction. Environmental and ecological certifications, such as EDGE (Excellence in Design for Greater Efficiencies), were also sought and ultimately achieved. It was recommended that the infrastructure include easy-to-find parking for people with disabilities, as well as bicycle parking facilities. The Central Government, through MINAM, arranged for the placement of 3,900 separate containers at all venues and coordinated with formalized recycling associations for the collection of recyclable waste, including paper, cardboard, plastic, and glass. Water dispensers were placed at the venues to encourage the public to use refillable water bottles and reduce the use of plastic bottles.

Among the key conclusions regarding food safety and hygiene issues is the importance of ensuring that packaging is reusable or made from recycled materials. Provide diverse food options and a varied, nutritious food offering as close as possible to the venue. This will prevent spectators from leaving the venue or traveling to get food. These types of sporting events require a lot of energy, including the use of lighting, television, radio, and internet broadcasting equipment, as well as air conditioning, among others. The use of renewable and low-carbon energy must be considered, and a crucial detail is ensuring that equipment, accessories, and lighting are turned off when not in use. What else was done during the games? More than 8000 trees were planted at the venues, and green spaces were created as part of the city's environmental legacy.

Once the Games were over, the project assumed new importance in order to leave a lasting sporting legacy in the city. With the administration of the main venues, such as Videna, Villa El Salvador, Villa María del Triunfo, Costa Verde, and Punta Hermosa, sustainability plans were developed that aligned with the sporting activities being held. Unfortunately, the pandemic postponed these plans, and the venues became vaccination centers that saved more than one life. With the return to so-called "normalcy," venues reopened their facilities to sporting activities, in addition to top-level sporting events such as the World Junior Athletics Championships and the World Junior Artistic Swimming Championships, both held in 2024 at the Videna headquarters. As part of these activities, the Peruvian Sports Institute (IPD) hosted an International Congress, titled "Peru Sport Summit," with one of the discussions focusing on "Environmental Sustainability in Large-Scale Events." This discussion prompted this essay, as it touched on topics that are very common but that we often overlook.

Kaleet Cahuana, who was Environmental Manager for the Lima 2019 Pan American and Parapan American Games, noted that "every sporting event marks a milestone in both sports brands and regulatory elements, and since 2012, starting with the London Olympic Games, ISO 20121 has been adapted to event sustainability." The International Olympic Committee (IOC), as we initially noted, has placed sustainability as a key element of its agenda. Proof of this is its commitment to reducing its emissions by 50% by 2030. The remaining 50% will be offset through its Olympic Forest project, which restores degraded landscapes in the African Sahel region and involves the planting of more than 400,000 trees. Starting in 2030, all Olympic Games will be contractually obligated to minimize their carbon emissions and offset more than 100% of their residual emissions. During the discussion, Kaleet also mentioned that "any high-performance infrastructure built in the country for the 2027 Pan American Games is subject to and obligated to comply with environmental obligations." In this regard, it is also important to discuss the community's responsibility to itself. If you have infrastructure in a location with a swimming pool and your primary need is water, efficiency lies in having a water recirculation system.

The same applies to a sports field with artificial lighting. "You can play soccer, especially at night, because that's where people watch the game, and high-powered floodlights with a high wattage are used, but there's a sustainability factor because the ticket price covers that. However, if practices are held at night, it's more expensive because after 5 p.m., peak rates are charged, in addition to a power charge, which means costs increase." While we mentioned the topic of food above, I think it's also appropriate to revisit it when discussing the logistics of holding a press conference or a coffee break for guests. What would be suitable? Using food, fruits, and beverages from the region would promote the economic development of the area, which would also add a touristic and distinctive touch. Another part of the learning process is the relationship with brands that could sponsor an event. Brands seek to be socially responsible and are now also including the topic of environmental sustainability (the case of Toyota in Paris, which we'll discuss later). When it comes to brands, communication is a crucial aspect to consider. On the one hand, there's the work we do with the media, which can spread the word about our event and reinforce the theme of environmental stewardship. This encourages a change in negative behaviors by emphasizing these concepts through narratives and stories on social media. On the other hand, we need to have a dialogue with our stakeholders, municipalities, police, and other authorities, to stay aligned on what we want to achieve to make our event sustainable. Some of these examples, seen at Lima 2019 and projected by

Legado, also continued at Santiago 2023. This is how the sustainability strategy was launched, turning the Games into Chile's first carbon–neutral mega-sporting event.

5 The Contribution of Paris 2024 and the Mexican Grand Prix (F1)

Drawing on the experiences of the London 2012 and Rio 2016 Olympic Games—although we won't mention Tokyo, as it was practically a family event—Paris 2024 sought to be the most sustainable in history. Upon reviewing the reports, several key organizational points can be highlighted. There was a 50% reduction in emissions compared to the 2012 London and 2016 Rio Games. One of these was the innovative energy solutions, as all venues were powered by renewable energy. These connections were maintained after the Games, reducing the ecological footprint and leaving a legacy in the city. For those venues that required power, biofuel generators or batteries were used. Only two new venues were built. Ninety-five percent of the 35 venues were either existing or temporary, with designs that sought to reduce their impact on biodiversity. The Aquatics Center, one of the two new venues, was built using low-carbon construction methods. The stage featured prefabricated seats made from recycled plastics, and wood was used in its construction to reduce emissions. The Olympic medals also included recycled materials, each inlaid with a repurposed piece of the Eiffel Tower.

Regarding responsible eating, local, seasonal, and plant-based products with a lower carbon footprint were consumed. Those that weren't consumed were redistributed. Plastic use was reduced by half. Coca-Cola, as a sponsor, installed water and soda fountains in all venues. Additionally, all attendees were able to access the venue with reusable water bottles. Other details to keep in mind: The Olympic cauldron used electricity and LED lights to give the impression of being on fire, making it 300 times less polluting than if it had burned gas, explained Georgina Grenon, Director of Environmental Excellence for the Games. Regarding mobility, priority was given to bicycle use—with 1000 km of bicycle lanes available—and public transportation (buses, subways, and trains), facilitating the movement of the expected 15 million visitors. The organizers used Toyota vehicles—electric, hybrid, and hydrogen—to transport athletes and the Games family. This last point is interesting to compare with Euro 2024. A UEFA analysis indicated that fan transportation and travel remain the largest contributors to the carbon footprint of major sporting events. As a solution, carbon offset programs can help mitigate the environmental impact of team and fan travel.

According to an AP report, "This summer's Olympic and Paralympic Games generated 1.59 million tons of global warming carbon dioxide, from athletes' meals and the construction of their rooms to spectator flights and the energy used for the events." To close this section and focus on other conclusions, Luis Báez mentioned something that caught my attention, which pertains to communication aspects and refers to the rise of virtual experiences: "Streaming platforms and digital participation tools allow fans to participate in events remotely, reducing the need for physical travel." This is a fascinating aspect because little by little, remote, virtual, and interactive experiences will begin to be created that will allow fans without the financial means to experience being in a World Cup stadium or experiencing a Fan Fest. On the Formula 1 side, 30 innovative initiatives were implemented during the 2024 Mexican Grand Prix. "Sustainability is not a sprint, it's a marathon. You have to walk it and move it kilometer by kilometer," said Sandra Castañeda, Deputy Director of the Formula 1 GP CDMX.

The first pillar was circularity. 43.9 tons of inorganic waste were recycled, and paper use was significantly reduced thanks to the implementation of 55,000 digital tickets. 7021 kg of food and 5119 L of water were donated to vulnerable communities. Additionally, during the event, "zero waste" hydration was promoted using 100% recyclable and collectible aluminum cups, thanks to an agreement made with ARZYZ. A second pillar was linked to well-being. Healthy food was provided (314 general meals, 54 vegetarian meals, 33 vegan meals, and 123 gluten-free meals), as well as sun protection (60,000 free pieces) and hearing protection (400 earplugs for children and 33,500 for adults). Additionally, accessible areas for people with disabilities were created. And the third pillar was Net Zero. During the 2024 race, a mobility plan was implemented in collaboration with the Mexico City government, featuring 88 eco-efficient public transportation units (electric and low-emission), which benefited 7688 people by reducing vehicle use and mitigating CO_2 emissions.

6 Conclusions About a Sustainable Future

Every sporting event has three main stages. The first is construction and preparation, in this case, the construction of venues, the development of guidelines, and the entire strategic aspect of the event that will lay the foundations for its development. Then comes the entire operational phase, which involves various challenges across the different functional areas. For this, all the components involved in the event must be clear about their responsibilities and how they should be sustainable. The third stage is the legacy, which is everything the Games leave behind for the city and is a fundamental aspect. Looking ahead to Lima 2027, the city will face new challenges and new opportunities to remain a sustainable city. When the proposal was presented and supported to Panam Sport, it was indicated that only five new towers would be built in the Pan American Village to accommodate nearly 10,000 beds.

In addition, the construction of a water park in Callao is planned for one of the sports that did not have a venue built for Lima 2019: rowing. This sport was practiced in Albúfera de Medio Mundo, about 189 km from Lima, specifically on Vegueta Beach. This was and is one of the outstanding debts to this sport. For Lima 2027, the construction of a field for this sport is expected, and we analyzed a LinkedIn post by Sergio Viacava, president of the Peruvian Rowing Federation, in which he refers to the sustainability that rowing can bring: "Rowing and nature go hand in hand, so integrating sustainable practices not only protects the environment and promotes a better society, but is crucial to advancing toward sporting success and achieving a first Olympic medal."

Viacava emphasizes the importance of building the future venue with reclaimed wood or other sustainable materials. Furthermore, he emphasizes that waste management policies must be implemented at events, responsible tourism must be promoted, and green sponsors must be attracted. Let's keep in mind that there is a "Sustainable Development Report." This report measures countries' progress toward the Sustainable Development Goals established by the UN. Peru is currently ranked 64th. Lima 2027 is drawing ever closer, and its success will depend on the coherence of public policies, effective coordination with local and regional governments, and the systematic use of renewable energy in sporting events—factors that will allow for continued progress and improvement in this area. In this context, sustainability in sporting events must be understood as a strategic investment that generates longterm environmental, social, reputational, and economic value. Sport, in addition to reducing its environmental impact, acts as a powerful educational platform capable of normalizing sustainable behaviors across broad sectors of society. For Lima 2027 to consolidate its identity as a true laboratory of green innovation, it is essential to move beyond passive awareness-raising and transform fans into strategic actors within the sustainability model. This can be achieved through tangible incentives and reward systems, including commercial benefits and ticket discounts, that promote responsible practices such as sustainable mobility, proper waste segregation, and the consumption of local and low-impact options. In this way, environmental commitment can be transformed into an aspirational value embedded in popular culture, reaffirming sport's potential as a catalyst for change. In this scenario, there is indeed another game being played—and it is for the planet.

References

Baez RL (2024) Descarbonizando el deporte: desbloqueando oportunidades de transformación energética. https://www.2playbook.com/palabra-tecnico/descarbonizando-deporte-desbloqueando-oportunidades-transformacion-energetica_17484_102.html#:~:text=Cuando%20las%20organizaciones%20deportivas%20abanderan%20la%20sostenibilidad%2C,clim%C3%A1tica%20sin%20precedentes%2C%20marcada%20por%20temperaturas%20r%C3%A9cord

Breitbarth T, McCullough BP, Collins A, Gerke A, Herold DM (2023) Environmental matters in sport: sustainable research in the academy. Eur Sport Manag Quart 23(1):5–12. https://doi.org/10.1080/16184742.2022.2159482

Briese R (2001) Sustaining Sydney: the 'Green Olympics' from a global perspective. Urban Policy Res 19(1):25–42. https://doi.org/10.1080/08111140108727861

Brundtland GH (1987) Our common future—call for action. Environ Conser 14(4):291–294. https://doi.org/10.1017/S0376892900016805

CarExpert (2024) Tiny Toyota EV goes for gold with Olympic heavy lifting. https://www.carexpert.com.au/car-news/tiny-toyota-ev-goes-for-gold-with-olympic-heavy-lifting

Chalkley B, Essex S (1999) Sydney 2000: the 'Green Games'? Geography 84(4):299–307. https://doi.org/10.1080/20436564.1999.12452707

Cury R, Kennelly M, Howes M (2023) Environmental sustainability in sport: a systematic literature review. Eur Sport Manag Quart 23(1):13–37. https://doi.org/10.1080/16184742.2022.2126511

Dohlsten J, Barker-Ruchti N, Lindgren E-C (2021) Sustainable elite sport: Swedish athletes' voices of sustainability in athletics. Qualit Res Sport Exer Health 13(5):727–742. https://doi.org/10.1080/2159676X.2020.1778062

Edwards MB (2015) The role of sport in community capacity building: an examination of sport for development research and practice. Sport Manag Rev 18(1):6–19. https://doi.org/10.1016/j.smr.2013.08.008

Forest Green Rovers (2025) Anothr way. https://www.fgr.co.uk/another-way/

Fuchs M, Hovemann G (2022) The circular economy concept in the outdoor sporting goods industry: challenges and enablers of current practices among brands and retailers. Sustainability 14(13):7771. https://doi.org/10.3390/su14137771

Giorgio P, Argyro-Elisavet M, Rita M, Francesca V (2025) Leading by example: how National Sports Federations implement environmentally sustainable change through human resource management. Leisure Stud 45:1–11. https://doi.org/10.1080/02614367.2025.2495272

Hada JS (2025) Utilizing Textile waste in the production of nanotechnology-based sports Textiles. In: Nanotechnology-assisted recycling of Textile waste, pp 285–309. https://doi.org/10.1002/9781394175017.ch10

Instituto Peruano del Deporte (2013) ¡Cuenta regresiva! Lima tomó oficialmente la posta para organizar los Juegos Panamericanos 2019. https://www.gob.pe/institucion/ipd/noticias/198272-cuenta-regresiva-lima-tomo-oficialmente-la-posta-para-organizar-los-juegos-panamericanos-2019

Kolyperas D, Maglaras G, Sparks L (2019) Sport fans' roles in value co-creation. Eur Sport Manag Quart 19(2):201–220. https://doi.org/10.1080/16184742.2018.1505925

Lau E, Hou H, Lai JHK, Edwards D, Chileshe N (2021) User-centric analytic approach to evaluate the performance of sports facilities: a study of swimming pools. J Build Eng 44:102951. https://doi.org/10.1016/j.jobe.2021.102951

Mani A, Manoharan G (2026) Green entrepreneurship: a road to sustainable entrepreneurial growth. In: Mehra P, Kumar P (eds) Incentives and benefits for adopting green entrepreneurship practices. IGI Global Scientific Publishing, pp 169–190. https://doi.org/10.4018/979-8-3693-9306-2.ch007

McMillan C (2024) 'Seeing it as a whole': a research agenda for the sociology of sport and climate change. Int Rev Sociol Sport 60(3):401–417. https://doi.org/10.1177/10126902241268182

Memari Z, Rezaei Pandari A, Ehsani M, Mahmudi S (2020) Business management in the football industry from a supply chain management perspective. Int J Sports Market Spons 22(4):737–763. https://doi.org/10.1108/IJSMS-02-2020-0025

Millington R, Giles AR, van Luijk N, Hayhurst LMC (2021) Sport for sustainability? The extractives industry, sport, and sustainable development. J Sport Soc 46(3):293–317. https://doi.org/10.1177/0193723521991413

MINAM (2014) Juegos Panamericanos de Lima 2019 se realizará cumpliendo estándares ambientales. https://www.minam.gob.pe/notas-de-prensa/juegos-panamericanos-de-lima-2019-se-realizara-cumpliendo-estandares-ambientales/

Morfoulaki M, Myrovali G, Kotoula K-M, Karagiorgos T, Alexandris K (2023) Sport tourism as driving force for destinations' sustainability. Sustainability 15(3):2445. https://doi.org/10.3390/su15032445

Müller M, Wolfe SD, Gaffney C, Gogishvili D, Hug M, Leick A (2021) An evaluation of the sustainability of the Olympic Games. Nat Sustain 4(4):340–348. https://doi.org/10.1038/s41893-021-00696-5

Olympic (1999) IOC guide on sport, environment and sustainable. https://stillmed.olympic.org/media/Document%20Library/OlympicOrg/Documents/Olympism-in-Action/Environment/IOC-Guide-on-Sport-Environment-and-Sustainable-Development-Table-of-Contents.pdf

Olympic (2018) Recycling project for the preliminary draw of the 2018 fifa world cup. https://stillmed.olympics.com/media/Document%20Library/OlympicOrg/IOC/What-We-Do/celebrate-olympic-games/Sustainability/Case-Studies/2016/IF%20SCS%20-%20FIFA%20Recycling%20Project%20for%20the%202018%20World%20Cup.pdf

Orr M, Inoue Y, Seymour R, Dingle G (2022) Impacts of climate change on organized sport: a scoping review. Wires Clim Change 13(3):e760. https://doi.org/10.1002/wcc.760

Pelcher J, McCullough BP, Trendafilova S (2020) Collegiate athletics environmental sustainability efforts within STARS reporting. Int J Sustain Higher Educ 22(2):328–343. https://doi.org/10.1108/IJSHE-07-2020-0246

Räikkönen T, Hedman J (2025) Unlocking the power of sports: an exploration of the nexus between shared place, community competence, and sense of community. Int J Sport Policy Polit 17(1):81–99. https://doi.org/10.1080/19406940.2024.2396836

Ren X, Ren Y (2024) Public environmental concern and corporate ESG performance. Finan Res Lett 61:104991. https://doi.org/10.1016/j.frl.2024.104991

Ross WJ, Pfahl M, Trendafilova S (2024) Racing with a purpose: sustainability in formula E. Commun Sport 13(3):534–559. https://doi.org/10.1177/21674795241233558

Schleifer R, Tamir I (2023) Sport, the arts, and fans' loyalty: the role of color for sport fans. Front Psychol 14:85. https://doi.org/10.3389/fpsyg.2023.1239085

Schneider S, Mücke H-G (2024) Sport and climate change—how will climate change affect sport? German J Exer Sport Res 54(1):12–20. https://doi.org/10.1007/s12662-021-00786-8

Song L, Kong X, Cheng P (2024) Supply-demand matching assessment of the public service facilities in 15-minute community life circle based on residents' behaviors. Cities 144:104637. https://doi.org/10.1016/j.cities.2023.104637

Subie A, Mouritz A, Troynikov O (2009) Sustainable design and environmental impact of materials in sports products. Sports Technol 2(3–4):67–79. https://doi.org/10.1080/19346182.2009.9648504

United Nations Climate Change (1992) United Nations conference on environment and development, Rio de Janeiro, Brazil. https://www.un.org/en/conferences/environment/rio1992

United Nations Climate Change (2023) Creating the greenest football club in the world: forest green rovers|United Kingdom. https://unfccc.int/climate-action/momentum-for-change/climate-neutral-now/creating-the-greenest-football-club-in-the-world-forest-green-rovers

Wahab S, Imran M, Ahmed B, Rahim S, Hassan T (2024) Navigating environmental concerns: unveiling the role of economic growth, trade, resources and institutional quality on greenhouse gas emissions in OECD countries. J Clean Product 434:139851. https://doi.org/10.1016/j.jclepro.2023.139851

Wang Y, Zhao Z, Shi M, Liu J, Tan Z (2024) Public environmental concern, government environmental regulation and urban carbon emission reduction—analyzing the regulating role of green finance and industrial agglomeration. Sci Total Environ 924:171549. https://doi.org/10.1016/j.scitotenv.2024.171549

Zhang D, Ortiz MA, Bluyssen PM (2022) A review on indoor environmental quality in sports facilities: indoor air quality and ventilation during a pandemic. Indoor Built Environ 32(5):831–851. https://doi.org/10.1177/1420326X221145862

Zhu R, Lin B (2026) How does technological innovation impact the convergence effect of energy and environmental efficiency? Empirical evidence from the mining industry. Environ Impact Assess Rev 116:108095. https://doi.org/10.1016/j.eiar.2025.108095

MIX
Papier aus verantwortungsvollen Quellen
Paper from responsible sources
FSC® C105338

If you have any concerns about our products,
you can contact us on
ProductSafety@springernature.com

In case Publisher is established outside the EU,
the EU authorized representative is:
Springer Nature Customer Service Center GmbH
Europaplatz 3, 69115 Heidelberg, Germany

Printed by Libri Plureos GmbH
in Hamburg, Germany